CONQUERING FAT LOGIC

Dr Nadja Hermann is a behavioural therapist with a background in nutritional science. She meticulously documented her own weight loss, researched countless scientific studies, drew cartoons to illustrate what she learned, and initially self-published an ebook in which she debunked all the old lies about losing weight. Along with the publication of her ebook, she also launched her blog fettlogik.wordpress.com, which has attracted many hundreds of followers. She lives with her husband in southern Germany.

NADJA HERMANN

CONQUERING
FAT LOGIC

*how to overcome
what we tell ourselves
about diets, weight,
and metabolism*

SCRIBE
Melbourne • London

Scribe Publications
2 John St, Clerkenwell, London, WC1N 2ES, United Kingdom
18–20 Edward St, Brunswick, Victoria 3065, Australia
3754 Pleasant Ave, Suite 100, Minneapolis, Minnesota 55409, USA

First published in English by Scribe 2019

The contents of this book have been carefully researched and
checked on the basis of sources considered trustworthy by the
author. However, this book should not be regarded as a substitute
for individual medical consultations. Please contact a qualified
doctor for medical advice. The author accepts no responsibility for
any negative effects which may arise in direct or indirect connection
with this book.

Printed and bound in the UK by CPI Group (UK) Ltd, Croydon
CR0 4YY

Scribe Publications is committed to the sustainable use of natural
resources and the use of paper products made responsibly from
those resources.

9781911617365 (UK edition)
9781947534711 (US edition)
9781925713206 (Australian edition)
9781925693300 (ebook)

CiP records for this title are available from the National Library of
Australia and the British Library.

scribepublications.com
scribepublications.co.uk
scribepublications.com.au

Contents

Foreword

As far back as I can remember, I was always overweight. Even as a child, I felt like a big lumbering oaf next to my three skinny stepsisters. It was so unfair, the way they were able to polish off huge portions of pizza or ice cream and not worry about their weight. When they came to visit and would stand giggling on the scales, not even knowing the significance of the numbers on its dial —'Fifty kilos, is that a lot?'— I would be painfully aware of the fact that I was almost twice that weight, even at the age of fifteen.

The difference was like night and day. While they came from a genetically thin family, my parents were morbidly obese — in non-medical language, grossly overweight — just like three of my grandparents (my maternal grandmother was 'only' obese — in other words, just fat). In my teenage years alone, I tried the Hay diet, the Atkins diet, the Hollywood diet, fasting, and more in my battle to become less overweight. I repeatedly lost up to 15 kg, but the 'yo-yo effect' would kick in and I would soon pile that weight back on.

When I was twenty, I weighed more than 130 kg. Then I went on a crash diet and in just a few months starved myself down to a weight of 68 kg, with a body height of 175 cm. It was the first time in my life that I had been in the normal weight range for my size. It didn't last long. My metabolism had slowed down completely and as a result

I just kept gaining weight again. When I was eventually diagnosed with hypothyroidism, I concluded that so-called normal weight just wasn't realistic for me. In my case, it would mean a life of permanent hunger and self-torture. After some reading, I found that such a life wasn't necessary: excess weight was demonised without reason. I might be fat, but I didn't smoke, drink, consume fast food, or eat red meat. And I was physically fit, apart from my excess weight.

I decided to set other priorities in my life — I completed my doctorate in psychology, trained as a psychotherapist, got married, and started renovating an old house. At the age of thirty, I tipped the scales at 150 kg, but was not limited in life by my weight. In short, I was happy with my conscious decision to enjoy life rather than leading an existence of constant hunger and self-denial.

I was open about my weight and, in general, it was not an issue for my husband, my friends, or the people I encountered through work. If asked, I said I was comfortable with my weight and that my only wish, if anything, was to get a little more active and be a bit fitter. But at that same time, I was visiting an obesity clinic as an outpatient and enquiring about a stomach stapling operation. I kept this fact secret from everyone because I didn't want to face the questions it would have prompted about my claim to be 'comfortable' with my weight. For various reasons, I eventually decided not to have the operation, and instead I buried myself even deeper in studies that showed that being overweight was not really harmful.

My doctor never mentioned my weight, but I rarely went to the doctor's anyway. Not because I was never sick, but because I didn't want to be confronted with the problem of my weight. Every time I did go, my blood pressure was enormously high, but I dismissed it as 'white-coat hypertension'. In fact, my blood pressure was just as high at home. To reduce it, I stopped taking the pill and started drinking a litre and a half of green tea every day, having read research that said that both things would help lower my blood pressure. And actually, it did go down slightly. It was still far too high, but I was relieved at the improvement I saw. I managed to ignore the fact that I was suffering frequent back pain and that I was having trouble sleeping, in part because of my heavy snoring.

This went on until one day I slipped while doing housework and injured my knee. I know now that I tore my cruciate ligament, but at the time my doctor said it was probably nothing to worry about and prescribed me ibuprofen. And it's true that after a week I was able to walk again without pain. From then on, I would sometimes feel a twinge in my knee, and it was susceptible to spraining, which would hurt for a few days, but the pain was tolerable.

Then I had another accident while renovating our house. This time, I ruptured my meniscus. Again, my doctor told me it was probably nothing serious, gave me ibuprofen and … no, this time it did not get better within a week. I walked with a limp for months, but my knee improved enough for me to be able to function in day-to-day life. After six months I must have pulled my meniscus or torn it again because I was laid low for another several weeks, unable to move. Six months later, the same thing happened again.

That last time was the straw that broke the camel's back. After more than a year of pain and restricted mobility, I had a breakdown. I realised that, over the previous few years, my health had decreased while my weight had continuously increased. And I knew that if I kept going this way, within a few years I would be severely morbidly obese and unable to walk — and still in my early thirties!

Something had to change. For the first time in my life, I consciously started thinking about my eating behaviour and began reading research on genetics, metabolism, diets, and obesity. Ironically, I was not a novice in this field. I'd graduated from a high school specialising in nutritional science, and my dissertation in psychology focused on diets. However, in my private life, although I followed the topic, I did so selectively — looking only at what I wanted to see. Now, I began to explore the 95 per cent of the research that I had turned a blind eye to before.

For months, I read everything I could get my hands on about obesity, metabolism, nutrition, and weight loss: articles, studies, forums, books, reports. Gradually, this led me to throw my previous 'knowledge' — all my *fat logic* — overboard.

I came across the term 'fat logic' on a website somewhere on the internet, and it immediately resonated with me, as it seemed to describe my lifelong beliefs until that point. The term doesn't mean 'fat people's logic' but refers to the complex grab bag of supposedly medical facts, well-meaning advice, homegrown ideas, and fantasies that make losing weight not only difficult, but actually impossible.

I don't consider myself stupid or naïve. I have always been the kind of person who questions things, I have an interest in science, and I have a doctorate. But still I believed in so much fat logic — probably because I was always surrounded by it, because I was told from an early age that our family had 'fat genes', and because it appeared to be corroborated by my own personal experiences. Tearing down the fallacies I had believed for my whole life was a long and sometimes painful process.

In the following year, I went beyond simply gathering theoretical knowledge, and began to put it to practical use. By September 2014, I was within the normal weight range, and a few months later I weighed 63 kg — the least I had weighed since the age of about twelve or thirteen. In 2014 I also started drawing occasional cartoons, which I published on my blog, illustrating my new discoveries and describing fat logic in general. The effect was — to put it mildly — polarising.

My readers didn't know anything about my private life, and some of them wrote irate comments, accusing me of having something against fat people. Though I lost a few followers, I gained many more new ones, and regularly received feedback from readers telling me that my blog and cartoons had helped them to recognise their own fat logic and change it.

But I began to realise that blog entries and cartoons were not the ideal media for communicating what I had to say. I decided to stop trying to pack all my hard-won knowledge into short cartoon strips, and instead put it all together in a structured way and make a book. The final product is a mixture of personal experiences and scientific facts. It has been, and remains, important to me to back up all my claims with academic research and facts — as a counterweight to fat logic, which is so resistant to change even when it's faced with hard facts. My main goal is to expose the system of supposed wisdom and well-meant advice about obesity and weight loss as nothing but fat logic, as a vicious circle of fallacies and myths. I intend to do it by taking each individual claim of fat logic that together make up that system and directly debunking them, one by one.

For me, breaking that cycle was the pivotal change. As a behavioural therapist, I failed to deal with my own weight for years. The reason is that (behavioural) therapy can only work if we recognise which behaviours we need to change, and we are convinced that therapy will be effective and we will be able to make it work.

Fat logic stands in the way of that recognition because it obfuscates the facts, and it includes so much contradictory advice that permanent weight loss comes to seem like an unattainable goal. Overcoming fat logic is the first and most important step towards achieving a body weight that we are actually comfortable with.

Whether and when that happens is a decision for each individual to make, and even if this book offers certain possibilities, it should not be taken as a self-help dieting book with a 'Follow these steps to lose 40 kg in two weeks!!' formula. It should, however, help readers to find their own path and gain some clarity in the chaos propagated by diet books, magazines, and some so-called professionals.

But this book is not just for people who want to lose weight. Its intended audience also includes people whose weight is within or below the normal range, and those who are either generally interested in the topic or have problems in similar areas. The final chapter is aimed directly at normal-sized people, and most chapters will be useful to those who want to gain weight as well as those who want to lose it. Fat logic is not just a problem for fat people; it's a general phenomenon. I have never yet met a person who was completely free of it. Some slim or athletic people are more immersed in fat logic than other severely overweight people who are constantly confronted with the issue — whether they like it or not. And even I sometimes discover some remnants of fat logic in my own thought processes and beliefs.

For example, a few months ago I was talking to a friend of mine who is a physiotherapist, and complaining that my ribcage is too broad, and that my ribs stick out. I went on for some time bemoaning my big bones and speculating that my stocky structure might mean that my body wasn't suited to normal weight. My friend's offhand remark, 'Just wait a couple more months,' really put me off my stride, especially when she followed up by explaining that the fat in my upper body had naturally pushed out my ribs and splayed them — it would just take some time before they returned to the normal position and then my chest would get narrower by a few centimetres. I was astounded; it had never occurred to me to doubt that I was naturally 'big-boned' — one of the typical fallacies of fat logic.

Each of the following chapters covers one such fat-logic fallacy, which enjoys widespread credence among overweight people, and which ultimately serves only to make losing weight more difficult. Some of the examples I use are gathered from day-to-day life or from forums, but most are beliefs I myself held steadfastly for years. Some of these facts and research findings will be difficult to believe, simply because we accept many fairytales about obesity and weight loss so unquestioningly as facts — some prime examples are the 'yo-yo effect', the body's 'starvation mode', and our social ideal of the slender body in recent years. If this book achieves its goal, reading it will open up new perspectives and incentives, and give readers a clear view of fat logic and how it stands in the way of achieving a healthy body weight and life style.

I only eat 1000 kcal per day, but I don't lose weight

There's one thing we can all agree on: everyone's body needs energy to exist. The question of how much energy it needs is a little more complicated. A widespread fallacy is that there are a huge range of differences in people's metabolic rates. The amount of energy we need is influenced by various factors, but the main ones are body mass, and what that mass is made up of.

A studio apartment requires less energy to heat than a four-bedroom house. The same is true of smaller and bigger people. A person with less body mass needs less energy to supply it. Furthermore, muscle mass requires more energy to maintain it than fat mass. This can be compared to the furnishings of a home: muscle mass is like the electronic devices that use up energy when they are on, but also use up some energy when they are in stand-by mode. Fat mass, on the other hand, just needs to be kept warm and supplied with blood.

Because of the biology of the human body, women naturally have a higher proportion of fat and a lower proportion of muscle mass than men. This explains why a man and a woman of the same size and weight will still have different energy requirements.

So, when we look at a scale of the human body's energy requirements we will see at one end the lowest requirement

— that of a small woman whose body weight is low (= low body mass) and who is extremely inactive (= little muscle mass, and little energy consumption due to movement) — and at the other end of the scale we see the highest requirement — a very big, heavy man (= high body mass) who leads a very active life (= a lot of muscle mass, and high energy consumption due to movement).

To put some numbers into this equation, a forty-year-old woman who is 150 cm tall and weighs 45 kg, and who leads an inactive life, will have a daily energy requirement of about 1200 kcal when doing nothing but lying in bed all day. At the other end of the scale, a twenty-year-old male body builder with a height of 200 cm and a weight of 100 kg will require 2500 kcal a day even if he doesn't move at all — that's how much it takes just to keep so much more body mass supplied with energy.

If we look at the normal daily routines of these two people (the inactive woman has a sedentary job in an office and spends her free time at home on the couch, while the body builder has a physical job that keeps him on his feet all day, and spends his evenings pumping iron at the gym), the difference is even greater. The woman will ultimately require around 1400 kcal a day, while the muscle man might use up more than 4000 kcal daily.

As mentioned before, these are extreme examples. Most people are somewhere between 150 cm and 200 cm tall, weigh more than 45 kg, and are neither completely inactive, nor top-level athletes. In a study carried out in 2004, Donahoo et al. (short for *et alii* — Latin meaning, more or less, 'and team') investigated the energy consumption of people while at rest — how much energy their bodies required if they were to just stay in bed all day. Around 96 per cent of participants were found to have a resting energy requirement within a 600 kcal range of difference, and 66 per cent had a resting energy within a 200 kcal range. This means that most people's resting energy requirements were neither extremely high nor extremely low, but within a relatively average range.

In 2005, Johnstone et al. studied 150 people and found that even those people with the lowest resting energy requirements still burned more than 1000 kcal in a day. This means that if such a person's calorie intake was restricted to 1000 kcal per day, they would necessarily lose weight. Even for our petite inactive woman, if she ate only 1000 kcal per day she would be eating 400 kcal less than she burned on a normal day. One kilo of fat mass is equivalent to about 7000 kcal, meaning that our small woman would shed about 2 kg a month on this diet.

It's even more dramatic if the woman in question is not petite, but overweight. The greater the body's weight, the more energy it consumes. For one thing, the body needs more muscle mass to transport 100 kg in weight from A to B than 50 kg, and for another thing, that greater body mass needs to be heated, as mentioned above. Thus, an overweight person has a higher basic energy requirement than someone of the same height but normal weight.

So, if the 150-cm-tall woman weighs 100 kg rather than 45 kg, her resting energy requirement will no longer be just 1200 kcal, but 1750 kcal per day (and if she engages in light physical activity, as much as 2000 kcal per day). If she consumes only 1000 kcal a day, she will already lose a whole kilo in weight per week. With a daily intake of 1000 kcal or less, weight loss would be inevitable in her case. For most people, a reduction in their daily calorie intake to below 1500 is enough for them to consistently shed several kilograms a month.

A person's energy consumption can be calculated relatively precisely using certain formulae. The only information you need is height, weight, sex, and approximate information about your daily activity levels. You can find plenty of online calculators on the internet; just search 'basal metabolic rate calculator' or 'BMR calculator' (your basal metabolic rate is the number of calories you would require if you were resting all day). Of course, these calculators are not 100 per cent accurate, but they can give you an estimation of your daily energy requirements with an accuracy of about 95 per cent. There's a very high probability that your BMR will lie somewhere between 1400 and 2000 kcal per day — unless, as mentioned above, you happen to fall into one of the two extremes of very high or very low body mass.

The bottom line is that most people consume far more than 1500 kcal per day, but even people with extremely low consumption still need significantly more energy than 1000 kcal. Which means it's practically impossible not to lose weight on a daily calorie intake of 1000 kcal.

But I really only eat 1000 kcal a day and still can't lose weight

Despite the common cliché of the fast-food-guzzling fat person, my favourite meal used to be a large mixed salad with pieces of salmon. I ate it regularly, and in my mental calorie journal I would estimate it contained about 500 kcal. When, after many years, I finally weighed out all the ingredients and calculated the actual number of calories they contained, I discovered that the dressing alone, with three tablespoons of olive oil, contained about 300 kcal. The number of calories in the salad itself — tomato, cucumber, red bell pepper, and lettuce — was within reason. Mozzarella, though, added considerably more to the total, and the fact that the salmon was fried meant that the final tally for this meal was 1500 kcal — three times the amount I had estimated, and equivalent to the entire daily energy requirements for a small, slim woman.

People can hugely misjudge their calorie intake, and overweight people have a strong tendency to underestimate the calorie content of their food. A study carried out by Lichtman et al. in 1992 investigated people described as 'diet resistant'. These people claimed not to be able to lose weight despite restricting their calorie intake to less than 1200 kcal per day. Measurements of their energy consumption showed that all the test subjects were within the predicted range (in other words, within the values estimated by a base metabolic rate calculator as described in the previous chapter) — which was, of course, well above 1200 kcal per day. However, in their nutrition journals, they *under*estimated their average calorie intake by 47 per cent (perhaps not as much as I did with my salad, but, still, well … a lot), and *over*estimated their physical activity by 51 per cent. They assumed their obesity was due to genetics or thyroid problems but perceived their eating habits to be normal.

Pietiläinen et al. carried out a similar study in 2010. This time, the test subjects were pairs of identical twins, one of whom was obese and the other of normal weight. The subjects were each asked to record their own and their twin's eating behaviour and physical activity. The scientists also took objective measurements of each subject's caloric intake. The results were interesting: self-reported eating and exercising in the subjects' food and physical activity diaries were approximately the same for both siblings in each set of twins. The obese co-twins believed, and reported, that they were eating and exercising the same amount and in the same way as their non-obese sibling. The co-twins who were of normal weight, however, reported that their obese siblings ate considerably more, and more unhealthily, and were significantly less active than themselves. The objective measurements confirmed the fact that the obese co-twins under-recorded their food intake by an average of 800 kcal and overestimated their energy consumption due to physical activity by around 450 kcal.

A study carried out by Burton (2006) shows that people are very bad at estimating the calorie content of food in general. In the study, estimates of low-calorie foods like chicken-breast and turkey sandwiches were found to be relatively accurate (around 500 kcal), but the higher the calorie content of a meal, the more inaccurate people's estimates were. On average, estimates of the calories in a meal actually containing 2000 kcal were just below 1000 kcal. A pasta dish containing 1500 kcal was estimated by test subjects at around 700 kcal. The extremely high-calorie dish 'cheese fries with ranch sauce', at 3000 kcal per portion, was estimated to contain only around 700 kcal. It seems that most people find it hard to imagine that a single meal portion can contain over 1000 kcal — just like I never imagined that there were 1500 kcal in my fish salad.

In a study of twenty-eight people of different weight categories carried out by Kroke et al. (1999), only one participant gave a close-to-accurate report of their caloric intake. Muhlheim et al. (1998) showed that people will continue to under-report their food intake even when they have been told in advance that researchers will be verifying their reports. They do count calories more carefully when they are told, as compared to when they are not told their reports will be verified, but they under-report all the same.

Even people who should know better, i.e., professional dietitians, are often unable to estimate caloric intake accurately. In their study, Champagne et al. (2002) compared the food intake estimates of an all-female group of dietitians and non-dietitians over the period of one week. The non-experts underestimated their caloric intake by an average of 429 kcal per day, and some estimates were even out by more than 1000 kcal. The registered dietitians' estimates were closer to the mark but were still on average

223 kcal per day too low, and some of the professionals got it wrong by up to 800 kcal. The fact that even experts with years of professional experience are unable to judge their caloric intake accurately is a clear sign that self-evaluation is just not a reliable exercise in this context.

The hard truth is that anyone who believes that they 'don't actually eat that much' but then still inexplicably puts on weight doesn't have a problem with their metabolism, but with their perception of their own eating habits.

Counting calories doesn't work for me

My slim friend eats way more than me

Overweight people are not the only ones who are bad at estimating their caloric intake; slim people seriously misjudge their calories, too. However, they tend to err in the opposite direction. One of the reasons for this seems to lie in our desired goals: people who want to lose weight tend to *under*estimate their food consumption, while those who want to gain weight, are more likely to *over*estimate it (Johansson et al., 1998).

I have often asked thin people who claim to eat huge amounts to weigh out their food precisely and record it. Funnily enough, just minutes before I sat down to write this chapter, the following message appeared in the comments section of my blog:

> I always thought I ate like crazy (and probably unhealthily), but when I stumbled upon this, I thought I'd try counting kcal. Sure enough, even when I bought a multipack of chocolate bars and ate them all within 2 hours, I was surprised to find that they contained far less than 8000 kJ (~2000 kcal). When I added up everything else I ate that day, I still only reached the normal daily requirement (actually a bit less than that, but I don't exercise much). Altogether, my energy balance is relatively normal — to my surprise — and even a little below the daily requirement. From my

personal experience, I can confirm that it's easy to overestimate what we eat.

So far, no one willing to do my little experiment has reported anything different.

A recent study by Kuhnle et al. (2015) yielded similar results when it investigated the sugar consumption of overweight and non-overweight people. It used both self-reporting diet diaries and objective measurements. The study found that those with the greatest objective sugar intake were, unsurprisingly, 54 per cent more likely to be overweight.

However, the exact opposite was true when it came to self-reported intake: those who reported consuming the most sugar were 44 per cent more likely *not* to be overweight. The slimmer subjects were generally of the opinion that they ate more sugar (and more in general) than their overweight peers, although in reality the precise opposite was true.

Another example of the distorted perception of underweight people is this one I came across in an online forum:

> Hello friends! I'm searching for people with the same problem as me. The problem being that I'm soooo skinny! I've always been able to eat whatever I want without putting on weight. I had a full medical check-up at my doctor's and everything is fine. He told me I just had to change my eating habits. Okay, fine, but after my gynaecologist asked me yesterday whether

I have an eating disorder, I realised something really needs to change. Today, I've been at home all day just eating: nuts, oatmeal, muesli … I've already got a stomach-ache and I've just worked out that I've still 'only' eaten 900 kcal. Is there anyone else out there who's trying to gain weight, and how do you do it? What do you eat? I'm soooo full, but I really want to gain weight and stop being so thin.

She can eat 'whatever she wants', but clearly that's not very much if she's only managed to consume 900 kcal after a whole day of desperately 'stuffing herself'. So, once again, we can see that people with small appetites perceive even a small amount of food as a lot — and they express that perception, too ('I've eaten soooo much, again!').

In the case of overweight people, this problem of perception could be at least partly explained by the fact that we often see others in situations where they are eating, for example in restaurants, at the cinema, at birthday parties, or at other festivities. These are situations in which even thin people often eat very large amounts. Research shows, for example, that the presence of other people increases the amount people eat by up to 72 per cent (Guyenet, 2014). This can quickly lead to the impression that other people eat far more than we do ourselves: while we often experience ourselves eating small meals, we only see others in social contexts in which people generally eat more.

Unlike overweight people, slim people often offset that increased consumption unconsciously and unintentionally by eating less in the following few days, simply because they feel less hungry than normal. Or a slim person will do considerably more exercise. Differences in between-meal snacking behaviour can also have a huge effect.

Of course, there are medical conditions that can lead to some of the nutrients ingested by an affected person not being absorbed and being excreted undigested (including fructose/lactose/gluten intolerance, diabetes, Crohn's disease, certain parasites, and a surgical shortening of part of the digestive system). In most cases, though, symptoms related to these conditions cause serious problems and so the conditions themselves rarely go undetected. (An extremely fast metabolism — or an overactive thyroid — is also such a condition, which I deal with separately later in this book. It is likewise associated with such serious symptoms that it would not normally remain undetected — and contrary to popular belief it doesn't cause an immense increase in calorie consumption.)

Ultimately, perception is the deciding factor: someone with a naturally small appetite and someone with a large appetite will perceive the same portion of food completely differently. Most of us know this from personal experience: a portion seems bigger when we don't really feel hungry, or if we don't like the food. The opposite is true when we are ravenously hungry, and wolf down a whole plate of food and only start to feel full after a second helping. This is exacerbated by the fact that the stomachs of people who regularly eat large portions become stretched in comparison to those of people who eat only small meals.

In most cases, a person who can 'eat anything she wants without gaining weight', simply *cannot* and/or *does not want to* eat more than she does, and she doesn't have the feeling she is denying herself anything. According to her perception, she always eats as much as she wants, while her overweight acquaintances complain of constant self-denial. Often, it is the occasions when she completely pigged out that lodge in her memory, while the fact that she forgot to eat breakfast, or that she didn't eat a thing during the eight hours between breakfast and dinner, are simply forgotten.

An overweight person, by contrast, will be more likely to underestimate portion size and to forget about the between-meal snacks they had. That piece of cheese from the promotion lady in the supermarket, or that handful of nuts offered by an office colleague can often be full of calories which add up through a day, but they are usually forgotten.

The only useful way to get a realistic idea of your eating habits is to weigh absolutely every mouthful with a set of kitchen scales and write it down. Estimating the calorie content of food is so inaccurate that it's as good as useless. This is even truer for people who see themselves as too fat or too thin and want to change their weight.

While I was writing this book, an article appeared on CBC News (2015) titled 'Diet research built on a "house of cards"? Nutrition studies depend on people telling the truth. But they don't.'

The article describes how for many years researchers gathered and published confusing or erroneous data based on self-report studies about nutrition. For decades, the article goes on, scientists have carried out research into the supposed paradox that obese people eat less than people of normal weight, leading them to investigate all kinds of theories to do with metabolism, genetics, or other possible explanations. But, CBC explains, scientists now realise that obese people actually eat more than people of normal weight and that their self-reported intake simply does not correspond to the objective data.

Archer et al. (2013) analysed the data gathered over decades from more than 60,000 respondents as part of a large-scale nutritional study in the United States. The researchers came to the conclusion that two-thirds of female respondents and nearly 60 per cent of males made physiologically impossible claims about their caloric intake. In the most recent survey, respondents of normal weight underestimated their calorie consumption by around 150 kcal, overweight respondents' estimates were around 180 kcal too low, and those of obese respondents were an average of 590 kcal below the real value. On the basis of those figures, it's easy to see how the impression could be created that overweight people eat less than people of normal weight.

The amount of confusion and mystification that surrounds this issue is actually absurd. The results of the Archer et al. study emphasise the huge importance of handling the results of nutrition studies with caution: how was the data gathered, and was it based on self-reporting or objective measurements? If the Kuhnle et al. study on sugar intake mentioned earlier had been based solely on questionnaires, it might well have appeared in the media under the headline 'Sugar makes you thin! Researchers discover that people who eat more sugar are more likely to be slim'.

That would be doubly wrong: because the data is wrong, but also because the media tend to confuse correlation with causation. In other words, they will use the headline 'Sugar makes you thin!' rather than 'Thin people report that they eat more sugar'.

I have recently been coming across many similar claims about chocolate — and googling 'chocolate makes you thin' will bring up no end of reports to that effect. Just one example is an article that appeared on *Spiegel Online* in 2012. Although it does point out that the relation between cause and effect is not clear, it says, 'The team led by Beatrice Golomb interviewed around a thousand people about their dietary habits [...]. In summary, it can be said that regular consumption of chocolate is associated with a lower BMI.'

The operative word here is 'interviewed', so the information was apparently self-reported. Later in the article, the author puzzles over possible causes 'since the sweet-toothed respondents were not any more active and, on average, they even consumed more calories than other subjects'. The author goes on to speculate whether there could be substances in cocoa that could influence metabolism.

At the time of writing, the first page of Google search results comes up with the following list of headlines from various news providers, which appear to be based on the above study:

- 'New Study: Chocolate makes you thin'
- '... a new US study has found that eating chocolate regularly can help achieve a slim figure'

- 'According to a new US study, eating chocolate does not make you fat. On the contrary: regularly snacking on chocolate helps you lose weight.'
- 'Chocolate makes you thin — eating chocolate regularly keeps you slim.'

So, a person who didn't know that self-reported data on nutritional habits is, to put it bluntly, complete bullshit, would see all kinds of astounding research results on the subject and would almost be forced to conclude that the secret to staying svelte is to spend all day stuffing yourself full of high-calorie confectionery.

My metabolism is wrecked

If my memory serves me well, I went on my first diet at the age of twelve. It was supposed to repair my wrecked metabolism and finally give me a faster/normal metabolic rate, which would mean I could eat whatever I wanted without putting on weight.

But my metabolism resisted any attempts to repair it — perhaps because it was never broken in the first place.

We often have the impression that our metabolism is some kind of capricious goblin inside our body, burning up calories, or hoarding everything he gets a hold of, as the mood takes him. In reality, our metabolism is the sum total of all the chemical processes in our body, including our brain function, heartbeat, the functions of our other organs, temperature regulation and, of course, muscle activity. When metabolism really shuts down, the symptoms are extreme. I personally suffer from Hashimoto's disease — an autoimmune condition which leads to thyroid insufficiency.

While I was losing weight, I also briefly had the symptoms of an overactive thyroid, due to the dosage of my medication not being adapted to my lower body weight. Both were extreme. When my thyroid was underactive, I was lethargic, slept excessive amounts but was still exhausted, was downcast — to the point of mild to moderate depression — and constantly felt cold and weak.

The symptoms of an overactive thyroid were even worse. I often woke up at night with my heart racing uncontrollably, I was extremely irritable, and I felt tense all the time. I barely slept but was filled with a feeling of nervous unrest and hyperactivity during the day. My pulse rate and blood pressure were far too high — and did I already mention the irritability? One false word and I felt like strangling the person I was talking to — a real advantage for a practising psychologist! However, despite these extreme symptoms, the difference in my caloric expenditure compared to my base level was ridiculously small — around 10 per cent, or approximately 200 kcal per day.

Al-Adsani et al. (2013) investigated the effects of thyroid hormones by inducing a light thyroid underactivity or overactivity in patients with the aid of medication. The difference in resting energy expenditure between thyroid overactivity and underactivity was around 15 per cent. That means that a person with a resting energy expenditure of 1500 kcal will burn a little over 110 kcal less if their thyroid is underactive, and around 110 kcal more if they have an overactive thyroid. Of course, thyroid under- or overactivity can be much more extreme, and when the condition becomes serious, the symptoms that accompany it become equally severe. When that happens, the issue of calorie consumption fades into insignificance in the face of these other symptoms, including substantial dysfunction of the body's organs.

So, over a month, a slightly underactive thyroid could make a difference of half a kilo, and a severe thyroid deficiency could make a difference of a whole kilo if it is not accompanied by a change in eating habits — but that's far less than you might expect given the very serious symptoms that would accompany that weight gain.

In reality, metabolism *cannot* be slowed down or sped up at will, and the largest factors influencing energy consumption remain muscle mass and physical activity. A study carried out by Toubro et al. (2013) found that 82 per cent of a body's resting energy needs were correlated with non-fat mass (muscles, organs), while a mere 10 per cent of calorie consumption was influenced by factors such as spontaneous physical activity ('fidgeting'), fat mass, or thyroid and other hormones. Our organs require a certain amount of energy to function. Over thousands of years of evolution, the body has optimised these processes so that they don't waste masses of energy — energy which the body can then easily store for times when the need might suddenly arise.

The rate of energy consumption is not slowed significantly until the body's fat reserves are reduced to a life-threatening level. A body must be severely underweight before its organs begin to fail, and only then does the rate of energy consumption become markedly reduced.

The oft-cited 'starvation mode' which the body is supposed to enter when denied calories is a myth. Weight-loss forums are full of comments like 'Dear God, whatever you do, don't eat less than [enter any number above 1000 kcal] or your body will go into starvation mode and store everything it can get!!!' That, if you'll pardon my saying so, is complete rubbish.

For the first six months of my weight-loss process, I couldn't run because of my injured knee. During that half-year period, in order to reduce the burden on my knee without the option of exercise, I cut my energy intake to 500 kcal a day. And what happened? … Nothing. Well, nothing except that I lost the amount of weight that I'd expected/calculated on the basis of that deficit. At the beginning, that was about 2 kg per week. Of course, as time went on, the rate at which I lost weight became a little lower, simply because the amount of energy my body required in the resting state became less as my mass went down.

As already explained in the first chapter, a larger/heavier body consumes more energy than a smaller/lighter one, and this means that during the weight-loss process, the body's energy consumption gradually sinks to a normal level, as corresponds to a person of normal weight. But that doesn't mean that the body's metabolism has become *slower*, it just means that it is gradually approaching the normal level of energy consumption of people who are not overweight. Of course, a radical crash diet can mean that the body receives insufficient quantities of important nutrients — mainly proteins and vitamins — and when that happens, the body will start to break down more muscle mass than necessary, which then further reduces its energy consumption. This is a nutritional issue and it is not an inevitable part of dieting, even if you radically reduce your calorie intake. However, whether your caloric deficit is large or small, it's always a good idea to undergo regular blood tests in order to make sure that your body isn't lacking any essential nutrients.

Incidentally, at the time of writing this chapter, I have now increased my caloric intake back up to more than 1900 kcal a day and the resulting small energy deficit means I am slowly shedding the final couple of kilos (to reach my target weight). The much-cited yo-yo effect never eventuated because I did not suddenly start 'storing everything my body could get'.

My experience is underpinned by scientific research, by the way. Studies of a number of people who lost weight by reducing their caloric intake showed that their metabolism didn't slow down — it's true that their energy consumption did reduce, but only by the amount expected as they lost weight. Thus, their base metabolic rate was normal for their body weight at any given time (Jebb et al., 1991). Weinsier et al. (2000) also found that test subjects who lost more than 10 kg solely by reducing their calorie intake (no exercise) didn't experience a drop in their bodies' energy requirements beyond the amount to be expected due to reduced overall body mass after weight loss. Amatruda et al. (1993) similarly found no difference in the energy consumption of formerly obese women who had attained their ideal weight and those of ideal weight who had never been obese.

Redman et al. (2009) examined several groups of dieters and did in fact find a small reduction in bodily energy requirements that could *not* be attributed to reduced overall body mass — the figure was 51 kcal per day in dieters with small calorie deficits, and 83 kcal per day for large calorie deficits. But this small reduction was only detected in subjects who lost weight without engaging in any physical exercise at all. Those who had exercised while losing weight did not show this reduction.

Rosenbaum et al. (2008) examined subjects who had lost significant amounts of weight over a period of several weeks by means of a liquid-formula diet of 800 kcal per day. The study followed a group who had maintained their reduced weight for a year, and another of people who had only recently completed the diet. The researchers compared these subjects with people of the same weight for whom that weight was usual and not achieved by previous dieting. The average resting energy expenditure of those who had never dieted was 1694 kcal per day. The subjects who had recently ended their 800-kcal-a-day diet used up around 1555 kcal per day; their daily energy requirement had thus fallen by some 139 kcal. The subjects who had completed their diet a year before had a resting energy requirement of around 1622 kcal per day.

It must also be noted here that the liquid-diet formula only contained 15 per cent protein, which means only about 30 g. This, for a fact, is less protein than the body requires and so, in my view, it's unclear whether the observed reduction in energy requirement was caused purely by the reduction in calorie intake and the resultant weight loss, or whether it was caused by a lack of essential nutrients. Whatever the case, the subjects' bodies obviously underwent a process of regeneration in the subsequent time period, as shown by the fact that the energy requirements of the group who had completed their diet a year before had risen once again.

The reason for minor reductions in energy expenditure is very likely to be the fact that dieters make fewer involuntary small movements — in other words, they *fidget* less. It seems that people of normal weight are more likely to compensate for increased caloric intake by *fidgeting* more. This is a behaviour I noticed in my slim stepfather. Often, while sitting at the dinner table, he would be constantly swinging his leg — something no one else in the family did, and in fact something which I'd never witnessed before. Since overweight people tend to become even less fidgety than otherwise while in the process of losing weight, it's highly likely that this is the explanation for small changes in energy expenditure during weight loss.

Another possible explanation is a slight reduction in body temperature. People in the process of losing weight without exercising are less active overall, including in their smaller movements. It's possible that, as a result, their bodies produce less heat, and that without that heat all bodily processes are somewhat slowed.

The thing is, if you look at the actual figures, it becomes clear that although these differences are distinct on a scientific level, they don't actually have a particularly significant effect in practice, when it comes to metabolism and weight loss. Even if a very radical calorie restriction without accompanying exercise leads to a reduction in a person's energy requirements of around 85 kcal, that doesn't mean the person has stopped losing weight. All it means is that very rapid weight loss slows down, and then only by less than half a kilo per month. Even in people who lose weight by means of unbalanced calorie reduction, including nutritional deficiencies, and without accompanying physical exercise, i.e., the most unhealthy and extreme method possible, energy expenditure does not sink vastly. Even under such circumstances, the human body can only cut around 10 per cent of its energy consumption. So even then, the reduction in the rate of weight loss is far less than 1 kg per month.

Interestingly, research carried out by Zauner et al. (2000) showed that the resting energy expenditure of people while fasting rose slightly in the first three days, before falling again. This plainly refutes claims that the body immediately enters some sort of 'starvation mode' or 'power-saving mode' after a couple of missed meals or a few hours without food. Those kinds of effects only kick in after a long period of time — if at all.

The kinds of remarks you'll commonly find in forums, and in general conversation, along the lines 'you aren't losing weight because you aren't eating enough!' are total nonsense. The greater the caloric deficit, the more fat the body has to burn. This process is not somehow U-shaped, as some people seem to believe. Their thinking goes something like this: 'If I eat fewer than 1200 kcal, my body will go into power-saving mode and I'll put on weight because my body will suddenly only require 1000 kcal. If I eat 1300 kcal, I'll lose weight because my body won't go into power-saving mode, and it requires 1700 kcal.' This 'logic' seems to be accepted as most likely true by many. My own experience may go some way to explaining the reason for this.

I always had more-or-less normal energy requirements for someone of my height and weight — I know this because I always lost exactly the amount of weight that you'd expect for my reduced food intake, even when I was only eating 500 kcal per day. I don't think my metabolism is especially great, particularly since I tick all the criteria usually associated with a 'bad metabolism' (genetic predisposition, lots of yo-yo diets in the past, underactive thyroid, polycystic ovary syndrome, etc.), and what's more I was basically immobile for the first six months of my diet. So, I am in no doubt that the crucial difference between me and a dieter whose energy consumption slows down really was the nutrients I was ingesting: I was very meticulous about making sure I ate enough protein, fats, vitamins, and minerals.

The fact that the studies that found a reduction in the energy consumption of dieters were the same ones that used fairly unbalanced diets, while other studies found no metabolic slowdown, is a clear sign, to me, that nutritional deficiencies were the reason for the lowered energy expenditure levels recorded in those experiments. Those kinds of deficiencies will throw a number of bodily functions out of kilter, impair the functioning of many organs, and can cause people to become less physically active and to reduce any unnecessary movements (fidgeting) to a minimum.

The whole 'starvation mode' theory doesn't make a lot of sense, as it would mean that our bodies only start to store fat when we are trying to lose weight — as if flying in the face of centuries of evolution and saying: *Energy? Meh! I don't need any of that.* But that would have meant that the very first famine faced by our ancestors would also have been their last.

It's fatuous to believe that our bodies have to 'learn' the principle of storing extra energy. From birth (actually, from before birth), our bodies begin building up energy reserves in the form of fatty tissue whenever they receive more than they need. Thus, the body could be seen as being in constant 'starvation mode', if that state is defined as one in which the body stores extra energy. Real starvation mode is the state the body enters when its proportion of fat drops below 5 per cent (a little higher for women).

The Minnesota Starvation Experiment is often held up as an example when starvation mode is discussed. In it, male volunteers were placed on a strict reduction diet of 1570 kcal — corresponding to around half their daily requirement while engaged in physical work. It is often claimed that starvation mode meant that the men stopped losing weight at some point, but that claim is nonsense. The volunteers continued to lose weight until their proportion of body fat reached dangerously low levels. If they had continued to starve themselves after that point, they would have met the same fate as people in times and areas of famine, or in concentration camps: their organs would have eventually failed, and they would have died.

This experiment is also often quoted as evidence of the dangers of calorie reduction, since the men suffered both physically and mentally during the study. The logic is, 'If these men displayed such extreme symptoms with a caloric intake of more than 1500 kcal, what could happen when you eat even less?' That argument is easy to rebut: firstly, 1570 kcal represented an enormous deficit for the men, since they were also required to carry out physical labour during the study. A male worker with a physically demanding job who has a daily energy intake requirement of more than 3000 kcal but only ingests 1570 kcal cannot be compared to a physically inactive female weighing 70 kg who works in an office and requires only 1700 kcal per day. Quite apart from the fact that the volunteers' deficit was very large, the experiment was not set up as a simulation of dieting but as a simulation of an actual starvation situation. The men were given food of the kind that was common in areas of famine at the time: very meagre, low in nutrients, and monotonous. It mainly consisted of potatoes and was completely unbalanced as it contained almost no protein or vitamins. The result was a group of men engaged in heavy labour who rapidly became malnourished and showed signs of protein and vitamin deficiencies. And still they continued to lose weight, even in these worst of all possible conditions, until their body fat reached life-threateningly low levels.

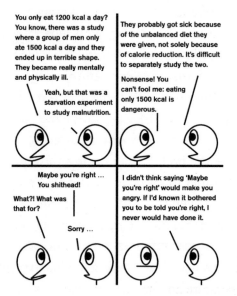

Another example goes back to 1965, when a man weighing more than 200 kg ate absolutely nothing for an entire year and slimmed down to 85 kg. Five years later, he had regained only seven kilos. This long-term fasting experiment was repeated in 1973 with a twenty-seven-year-old man. He fasted for 382 days and took only vitamins and sometimes mineral supplements. The patient lost around 125 kg and his doctors reported no negative effects on his health (Stewart & Fleming, 1973). This is certainly not the healthiest way to lose weight, and I am not, in quoting this example, encouraging anyone to emulate it. My concern is simply to illustrate that all the myths about the minimum numbers of calories that should be eaten are precisely that: myths.

Scientists now believe that, under normal circumstances, the body maintains the optimum running condition for all its processes to function normally. In cases of insufficiencies or deficiencies, the body seems to be able to make energy savings of a little under 10 per cent without it resulting in major symptoms. If reduction of energy consumption continues further — for example due to severe diseases of the thyroid gland that disrupt energy metabolism, or due to real starvation mode when body fat sinks to dangerously low levels — the body has to begin shutting down organ functions in order to save energy, and that results in severe symptoms and even becomes a danger to life.

The maximum amount of energy the body can save without severe symptoms appears to be only a couple of per cent, which it achieves by means of a slight reduction in body temperature and in small movements. The worst that can happen to your metabolism, then, is that you'll have the energy requirements of a person who is a few centimetres smaller than you. Even then, those not-quite-100 kcal that are not being consumed can be compensated for by, for example, increasing your muscle mass by 3 kg through physical training, which will increase your resting energy consumption by about 100 kcal. Alternatively, ten to fifteen minutes of exercise each day will also compensate for that 100 kcal change. However, those who are careful to take in enough nutrients while they lose weight, and who perhaps also do some strength training, will find the change in their body's energy requirements insignificant.

WHAT SCIENTISTS WRITE

In our study, the energy consumption of dieters fell gradually by 250–300 kcal. Some 240 kcal of that reduction was due to the reduction in body mass. The rest, 10–60 kcal, could not be explained by body mass, and we assume that the remaining reduction was due to a small drop in body temperature and a reduction in involuntary fidgeting movements. The effect was reversible after dieting was discontinued.

WHAT THE MEDIA MAKE OF IT

Beware the power-saving mode! How dieting destroys your metabolism. Researchers find losing weight makes your metabolism slow down by several hundred calories!

WHAT READERS UNDERSTAND

No wonder I've put on 8 kilos since I did that fast eight years ago! My metabolism is wrecked!!

In any case, the fact is that the fewer calories we eat, the more fat we burn. The more we reduce calories, and the less sport we do, the more likely the body is to adjust its energy consumption downwards slightly. But the fact still remains: the smaller the amount of energy you consume, the faster the body's energy stores are used up.

Eating too little is dangerous: you're left with no energy and reduced muscle mass

This is only half true. The fact is that most radical diets are not oriented towards sensible intake of nutrients but concentrate solely on ingesting as few calories as possible. Which is indeed dangerous. The body requires certain nutrients that it cannot manufacture itself and that must therefore be taken in with our food.

For example, the body needs protein to carry out necessary repairs, supply muscles, etc. A non-active person requires at least 0.8 g of protein per kilo of (normal) weight. That means for me, as a 175-cm-tall woman, around 60 g per day is the absolute minimum to keep me functioning. One gram of protein has 4 kcal, which means I need to eat 240 kcal worth of protein to avoid suffering from protein deficiency. As soon as physical exercise is involved, the amount of protein required by the body increases up to as much as 2 g per kilo of body weight (so, for me, around 150 g/600 kcal per day).

Protein deficiency is marked by symptoms such as exhaustion, circulatory problems, pallor, weakness, lack of concentration, depression/irritability, hair loss, and water retention, especially in the abdominal region. Many people experience these kinds of symptoms during a radical diet and they often make the mistake of attributing them to calorie reduction rather than lack of certain nutrients.

Of course, the more calories you take in, the more likely it is that you will also be getting enough protein, but even eating 3000 kcal a day is no guarantee of that. A protein-rich diet with fewer calories can contain more protein than that of someone who eats fruit for breakfast and pasta or rice for lunch every day and spends every evening snacking on crisps or chocolate. Such a person might polish off ample calories and still suffer from malnutrition. According to the Canadian newspaper *National Post* (2015) overweight patients in particular are often more likely to suffer from deficiencies of important vitamins, minerals, and proteins. Their diets tend to be high in calories, but unbalanced, leading to malnourishment despite their obesity. *Scroll* writes of a major problem with child obesity in India, with very overweight children showing such severe signs of nutritional deficiencies that their health, their development, and even their lives, are in danger. The reason for this is, of course, their diet, which is rich in calories but poor in essential nutrients.

During my period of extreme calorie reduction, my committed taking of sufficient protein and vitamin supplements, as well as undergoing regular blood tests, meant I was actually even able to redress previous deficits in certain nutrients (iron, vitamins D and B), and I had no health problems. I was sometimes asked whether I felt weak, faint, or similar, but I never did. And, compared to before my radical diet, my skin became significantly clearer and my hair stronger.

Although the quality of my diet actually improved despite radical calorie reduction, some of those around me took a different view. The general opinion of such a severe calorie reduction was that it was irresponsible, unhealthy — dangerous! This belief is lodged so firmly in people's minds that we are bombarded from all sides by the idea that everything is fine as long as we eat 'enough' (calories). What that means is that if I had added a couple of chocolate bars to my small, high-protein meals to raise my daily intake to 1500 kcal, observers would have considered my diet more healthy.

What people fail to understand is that a *caloric* deficit does not automatically mean an *energy* deficit. Fatty tissue is nothing *but* stored energy. It can be compared to a well-stocked pantry, where long-lasting basic foodstuffs like potatoes, pasta, flour, and canned and frozen foods are stored. More perishable foods still have to be bought in fresh, but the stockpile covers most nutritional needs. The fatty tissue in our bodies is sufficient to cover our basic energy needs, and we just need to add the vitamins, proteins, and a little of the fat that the body cannot keep in store.

Another very common prophecy of doom is that people who are losing weight are also supposedly in danger of losing muscle mass. Weight reduction is almost automatically equated with muscle loss, and men in particular often worry about losing strength and power when they lose weight. But, again, the rule here is that it is not caloric intake that is important, but the amount of nutrients in your diet (and how much you exercise). The body loses muscle mass for two reasons only: the muscle is not (no longer) needed; or the muscle can no longer be maintained. Muscle mass is important, but it also costs energy, and so our body only carries around as much as it needs — no more, but also no less. As long as the muscle is necessary, the body will not willingly break it down, seeing as it is the best investment for future survival and a guarantee of continued energy supply.

It would perhaps be logical in our underactive society to break down muscle in order to save energy, given that providing our bodies with energy involves no more than a trip to the supermarket or even a simple phone call. In the past, though, obtaining food was highly dependent on physical strength, speed, and stamina, and a hunter who immediately lost muscle mass during a period of famine would have had a smaller chance of hunting and killing the next animal. So, our bodies don't voluntarily start shedding muscle mass, but prefer to take energy from their fat reserves, which also happen to be much more efficient at storing energy.

Muscle mass and fat mass are separate systems and when one of them is built up or broken down, the other does not inevitably have to follow suit. Two things are necessary to maintain or build up muscle mass: the stimulus of physical training and sufficient nutrient supply. Training stimulus signals to the muscles that they are required by the body. When athletes stop training, their fitness level begins to deteriorate after just two weeks of inactivity. When our muscles are not used at their maximum strength, our bodies break down the unnecessary mass and adapt to the actual level of strain they are under. However, if we expose our bodies to stimulus from training while we're losing weight, the body recognises that requirement and adapts to it — if muscle mass is clearly required, then the body will keep it.

Nutrients play a crucial role here. The body requires nutrients both to maintain and to increase muscle mass — these are mainly proteins, vitamins, and minerals. When muscles are put under strain, tiny injuries can result — we feel them in the form of aching muscles after exercise. If no nutrients are available to repair those injuries and increase muscle mass, the body is left with no choice but to break down the damaged muscle. This means that, ironically, an exercise program which is not accompanied by a balanced diet can even lead to *more* muscle mass being broken down, since sport causes more tiny muscle injuries. Anorexics especially tend to exercise excessively as part of their illness, and, as their bodies are also undersupplied with nutrients, they tend to lose muscle mass by exercising.

In a 2014 study involving obese older males in a thirteen-week weight-loss program, Verreijen et al. showed that subjects could increase their muscle mass and strength with weight training and a sufficient supply of nutrients. They were compared to the members of the control group, who didn't exercise, and who lost weight but also lost a small amount of muscle mass. This clearly shows that reducing calorie intake does not automatically lead to a loss of muscle mass.

The bottom line is that our body can cover its pure energy needs by using up fat reserves. Fitness and performance are less dependent on the bare amount of food we eat than on the nutrients that food contains and the level of activity we maintain.

I only ate an apple a day and I *still* put on weight

Whenever I write or speak about weight issues, I find that a lot of people in my audience are unfamiliar with the concept of water retention. I believe this lack of knowledge is one of the problems when it comes to weight control.

A large percentage of your body is made up of water, and it is able to store several litres at short notice. Normal daily fluctuations in weight due to water can easily amount to plus or minus 3 kg, and considerably more in extreme cases (up to 5 kg or more).

One kilo of fat tissue corresponds to about 7000 kcal. A person of average height and weight burns around 2000 kcal a day, and a calorie reduction of 1000 kcal per day will lead to a loss of about 140 g of fat daily. That works out to about one kilo of fat lost per week. The problem is that this amount of weight loss is well within the range of natural fluctuation due to water. So a person can easily eat very little for three days and lose half a kilo of fat, but also retain a litre of water in that time. When she steps onto the scales, she'll think, 'What? I've barely eaten for three days and *gained* half a kilo?!'

On the other hand, it's fairly typical for people to experience losing several kilos in weight very rapidly at the start of a diet thanks to water loss. The carbohydrates our bodies store for short-term energy supply bind a large amount of water. The dieter's body will consume these energy stores first, before it starts burning fat reserves, which means a large amount of water is 'released'. However, that water is retained again by the body as soon as the dieter eats enough to restock those short-term stores. This is one of the origins of the myth of the so-called yo-yo effect.

Real weight loss can only be recorded by measuring weight over a long period of time, rather than from one day to the next. After a whole day of really pigging out, it could easily happen that your scales show a drop in weight, simply because your body has lost some extra fluid for hormonal or other reasons. Or vice versa. So one day means nothing in isolation.

Possible causes of small-scale fluid fluctuations include those listed below:

- fluid intake — it seems counterintuitive, but the *less* you drink, the *more* water your body retains;
- temperature — people can retain water in their legs and/or hands in hot conditions;
- hormones — many women retain water before their period (hence symptoms like swelling or tenderness in the breasts, bloated belly, etc.);
- physical exercise — physical exercise often causes tiny injuries to the muscles, which leads to swelling in those muscle areas;
- high salt intake — eating salty food causes short-term water retention;

- and sweating — saunas or intense physical exercise can cause short-term weight loss of a few kilos due to dehydration.

Larger fluid fluctuations may be due to the following:

- medications with a long-term hormonal influence, such as cortisone or the contraceptive pill;
- medical conditions such as hypothyroidism (if untreated);
- lack of nutrients such as protein or vitamin deficiencies;
- and food allergies/intolerances.

It is important to be clear that these kinds of fluctuations due to fluids are not *real* excess weight. Someone who gains a kilo due to fluid retention is no more overweight than someone who gets on the scales wearing a thick winter coat. It might be annoying, but it's not a reason to panic and suddenly start dieting. On the contrary: a diet in that circumstance could reduce healthy body fat values to unhealthily low levels, leading to being underweight when the fluid retention abates. One way of telling whether extra weight is because of fluid fluctuations or due to fat storage is by looking for pressure marks on the skin, for example from trouser waistbands or socks — fluid in the body's tissues leads to these marks. Very rapid weight gain is also a sign of water retention. A person who ate twice their normal daily amount would only put on around 300 g per day. So if you find yourself gaining several kilos within a week without having increased your eating considerably, the most probable cause is fluid retention. If you experience sudden, extreme water retention, you should be seen urgently by a doctor, who will test whether it is caused by illness or a lack of certain nutrients. This level of fluid retention definitely should not be treated with diuretics (drugs which increase the excretion of water) or by dietary means without medical advice.

In my view, what follows is a typical development of a diet.

Person X decides to finally start doing more exercise and eating less. In the first few days, her carbohydrate stores are used up, several litres of water are excreted, and the scales show a weight loss of 2 to 3 kg over the next few weeks. Person X is pleased — *It is possible! I'll keep up the good work!* — and continues to exercise and eat less. Physical exercise causes tiny rips in person X's muscles, increasing the body's need for protein in order to repair itself. But person X's diet includes mainly fruit and vegetables as they are seen as healthy. This gradually leads to a protein deficiency. This deficiency and the fluid retention it causes mask any loss of fat on the scales, and although person X is exercising and sticking to her diet, she stops losing weight, or even gains a little.

Now, along come all the well-meaning people who say things like, 'Well of course, your body's in starvation mode now, so it's holding on to everything it can get. You have to eat more if you want to lose weight!' So, person X starts eating more again, putting an end to her protein deficiency, and all that retained water is finally excreted from her body. Although she is now eating more, person X does not gain weight, and may even lose a little. *Aha!* she thinks, *I really was eating too little to lose weight.* But, having returned to her original eating habits, she slowly begins to gain weight again. 'Sure,' her friends tell her, 'starving your body has wrecked your metabolism. That's what you get for starving yourself.'

And so, person X is left with the impression that her starvation-metabolism makes her put weight on when she does not eat enough. And she also puts on weight when she eats more. No matter what she does, she can't lose weight.

Person X could have done any of the following, and she would have avoided this scenario:

a) She could have changed her eating habits less radically, with a smaller caloric deficit. Her weight would not have fluctuated so much, and she would have lost weight gradually.

b) She could have carried out her radical calorie reduction under medical supervision, with blood tests, vitamin supplements, and sufficient protein. She would not have had so much fluid retention, and her increased weight loss would have gone on more or less continually.

c) She could have continued her radical calorie reduction despite her stagnating weight loss. At some point her body would not have been able to retain any additional water. Despite her water retention, her weight would have gradually gone down due to the loss of fat.

Unfortunately, however, very few people follow any of these options. And importantly, option c) is not to be recommended, as prolonged nutrient deficiencies can cause damage to the body's organs, bones, and muscles.

Paradoxically, such a protein-deficient situation can even mean that well-intended physical exercise has the opposite to the desired effect and causes a *loss* of muscle mass. In principle, two conditions must be met for muscle mass to be built up: physical strain on the muscles so that they are stimulated to develop, and a sufficient supply of the right building materials, i.e., proteins. Placing strain on muscles causes tiny injuries which must be repaired. If there are no protein materials available for this, the body is forced to break down the damaged muscle tissue.

In this way, eating so little that the body no longer has sufficient supplies of protein and doing lots of well-intended physical exercise at the same time can actually lead to muscle mass being lost more quickly than without any sporting activity at all. It could be the case, then, that by reducing her intake of nutrients and doing lots of physical exercise, person X lost muscle mass and further reduced her energy needs after her diet.

I have personal experience of this protein problem, and I am glad to say that by the time it happened to me, I had already abandoned most of my fat logic and so managed to avoid falling into the typical dieter's trap. After about six months of only eating 500 kcal per day, my physical condition gradually began to improve enough that I was able to start doing physical exercise. I began with weight training and an exercise bike. Since I had lost weight continuously and radically during my calorie-reduced diet, I thought, *If I now burn 500 kcal a day by exercising, I should start losing weight even faster. Wonderful!* And that was indeed the case for the first few days. But then my weight loss began to level off. This is a graph of my weight loss over that twelve months:

Diet Report

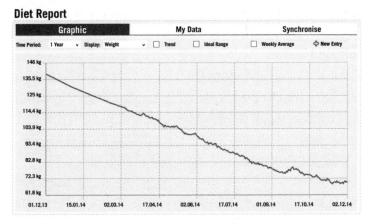

As you can see, on 19 April, at the start of my exercise program, my weight was 107.3 kg. Over the next four days, it rapidly went down to 105.2 kg. I was pleased with this, even though I knew that 2 kg in four days was a little bit too much, and that, realistically, about one kilo of that loss would be water. Still, I expected to keep on shedding weight rapidly. On 9 May, *sixteen days* later, and still eating only *500 kcal* per day and now also doing about an hour's exercise each day, my weight stood at 105.3 kg.

In the past, this was the stage at which I would have given up, convincing myself that my body must be in starvation mode if, even on so strict a program, I was failing to shed weight. But this time, rather than throwing in the towel, I took a blood test.

The test showed that my exercise program had increased my body's need for protein and that my food intake of 500 kcal a day was not enough to cover the increased requirement. Incidentally, this was the only time during my weight-loss process that I felt physically unwell and had problems with my circulation and hair loss. It only lasted about two weeks, but in that period I found a scarily large amount of hair in my hairbrush, my skin was dry and pale, and I remember feeling generally weak. For me, it was a real wake-up call about the effects a nutrient deficiency can have on the body and the importance of a balanced diet containing all necessary foodstuffs.

Together with my doctor, I decided to double my daily caloric intake and to make sure that it included 150 g of protein per day. Over the next few weeks, the retained fluid slowly began to leave my body, and I lost a great deal of weight — enough that by 18 May I weighed around 100 kg. The other symptoms I'd been suffering also quickly faded away, and I felt fit and well again.

The graph above shows that I lost weight continually over all, and, on average, at the calculated rate, but daily fluctuations of 1 to 2 kg were perfectly possible. It also shows that there were some periods of several days with no weight loss, and even some weight gain — even though my caloric intake remained constant.

This was an important eye-opener for me as in the past I'd always assumed I could quickly shed a few kilos if I just ate nothing for a couple of days. But what I hadn't realised was that those 'few kilos' were nothing but water, which my body would retain again immediately. With an approximate energy requirement of 2000 kcal, it is only physically possible for me to lose about 280 g of fat per day by not eating. Even if a day of fasting shows up on the scales as a short-term 'success' of two lost kilos, only around one-tenth of that amount is real weight loss.

Disease and medication can make you fat, and there's nothing you can do about it. (My doctor said so.)

I understand that doctors don't always have the time to advise patients in great depth about the effects of the medication they are taking. But oversimplified statements like 'may lead to weight gain' are read by the non-specialist public as 'You might put on weight from taking this. Don't worry about it if you do, and definitely don't bother trying to lose weight by dieting.' Propagating this kind of false information is irresponsible — especially when it's spread by the medical profession. In fact, medications usually only lead to weight gain in combination with other pre-existing causes, and those causes should be identified wherever possible because once they are known, patients certainly *can* do something to combat true weight gain.

The most serious illness which causes obesity is Prader-Willi Syndrome (PWS). The website of the German Prader-Willi Association (prader-willi.de) describes the condition as follows:

> People with PWS are usually small in stature. Their muscles tend to be very weak, especially in early childhood, which often makes physical activity

> difficult and tiring for them … People with PWS
> are unable to feel 'full'. A defect in the thalamus
> means they have no feeling of satiety, leaving them
> unable to regulate their eating behaviour.

This means that people with PWS have both a permanent uncontrollable appetite for food and also reduced muscle mass, meaning their food requirements are actually lower than normal. Combined, these are the worst possible conditions for avoiding obesity. Nevertheless, in a controlled environment, even people with PWS can achieve and maintain normal weight, as long at their energy intake is carefully monitored. Once again, it comes down to the balance of energy intake and output, although in people with PWS this balance is extremely difficult to maintain.

Another medical condition which can cause weight gain in women is polycystic ovary syndrome (PCOS), which causes hormonal imbalances. Even so, a survey by Norman et al. (2004) of past research showed that there was no relevant difference in rates of success in weight-loss programs between women with PCOS and women without it. The difference in their energy requirements was also very slight or non-existent. In one study which showed small differences in energy requirement, the possible weight gain due to metabolic rate was calculated to be less than 2 kg within a year — and even then, only if caloric intake was not altered to take the metabolic difference into account.

Then there is the always-mentioned hypothyroidism (underactive thyroid), which really does cause the body's metabolic rate to slow down. I am always embarrassed when people try to praise me by saying things like, 'It would have been so difficult for you to lose weight with an underactive thyroid condition. It's incredible that you did it!' It's embarrassing because it isn't extra-hard for me to lose weight. As long as my hypothyroidism is treated with hormone replacement tablets, there's no fundamental difference between me and a healthy dieter. Also, the metabolic deceleration is not all that big; at 10 to 15 per cent, you can certainly feel it in energy levels, but it can be compensated for by dietary changes or by 20 to 30 minutes of physical activity. Nonetheless, the condition does initially result in uncontrolled weight gain, which has nothing to do with food intake. On its website, the American Thyroid Association states that hypothyroid patients may experience fluid retention resulting in weight gain of approximately 2.5 to 5 kg. This is obviously a large initial increase, but it isn't real weight gain, as it can quickly be lost again with the right treatment. It will not cause obesity in anyone.

Fluid retention is often the cause of illness- or medication-related weight gain, especially if the medication contains hormones or cortisone. There's not much that can be done to counteract this kind of weight gain, so the best approach is to endure the effects patiently, secure in the knowledge that they will disappear again once the course of medical treatment is over. Although you can also talk to your doctor about tackling the fluid retention with treatments like lymphatic drainage or diuretic tablets.

I remember watching a daytime talk show some years

ago in which an incredibly arrogant thin woman said that she'd gained weight in the past due to the medication she was taking, and that she simply didn't understand why fat people couldn't lose weight. She had been able to lose 7 kg in two weeks without even trying. It is highly likely that this woman *didn't* have to try very hard to lose 7 kg — of water. It isn't comparable with the feat of losing 7 kg of fat (i.e., 49,000 kcal). To lose 7 kg of fat in two weeks, the talk-show lady would have had to shed half a kilo per day. If her daily energy consumption is estimated at a rather generous 2000 kcal, that would mean doing about three hours of intensive physical exercise every day while eating absolutely nothing. Although with that regime the chances are high of losing not only 7 kg of fat, but also her life — or at least of achieving total physical collapse.

Fluid retention is not the only way in which medication and illness can affect a person's weight. Some medicines make patients more energetic, others make them lethargic. The patient might not even be conscious of the effect, but these kinds of drug can lead to changes in your daily habits, reducing your general level of activity, and causing you to opt for the elevator rather than the stairs, choose to lie down rather than do some gardening, or to put your gym visit off till tomorrow. Over time, this can lead to a drop in energy consumption of a few hundred to a thousand calories per month.

A medicine's effect on the appetite should not be underestimated either. Very few people count every single calorie precisely or weigh out every single gram of food before they eat it. As we have already seen, we tend to be very bad at judging the amount we actually eat, and a change in appetite can also cause a change in this ability to perceive how much we are eating. It can lead people to pile a little more food on their plate or to reach for the snacks drawer in their desk more often, without giving them the feeling of having eaten more than usual. Their minimum for feeling that they have eaten enough increases along with their increasing feelings of hunger. What we normally notice is changes away from our *baseline* behaviour. If an altered appetite makes us feel permanently hungry, we may feel like we've eaten less than usual when in fact we have eaten more — simply because we feel hungry more frequently or we feel we have to fight feelings of hunger more often.

The bottom line is that medicine and illness can make it more difficult to achieve a balance between energy intake and expenditure. But they do not make you fat *per se*, nor do they remove your ability to lose weight.

Obesity is largely due to your genes

The reality is that the relationship between obesity and genes is similar to the relationship between obesity and medication or illness: genes create a basic situation, but they don't oblige anyone to be fat.

Things that really *can* be explained by genetics are appetite, preferences for certain flavours (such as sweet or fatty), and the natural urge to be physically active. Several studies (for example a German study carried out by Haupt et al., 2009) have shown that carriers of so-called obesity genes consume on average 125 to 280 kcal per day more and have *no* differences in their metabolic rates.

To say that some children have a genetic propensity towards obesity means only that they have an inherently larger appetite than naturally slim children, who feel hungry less often. But the deciding factor in whether children have a tendency towards being fat is the set of conditions created by their parents and the rest of their environment (school cafeteria meals, etc.), which can serve to either encourage or discourage obesity. Living in a household where high-calorie food is constantly available won't necessarily make children fat if their genetics mean they have a naturally small appetite. These children will simply have no desire to eat all the food they are offered. Children with naturally large appetites, by contrast, will pounce on the proffered fare. By the same token, those children will be unable to overeat in households where there are not a lot of high-calorie foods.

It is also possible to counterbalance a genetic pre-disposition through parenting — although it can also be reinforced through parenting. When a *naturally thin* child experiences being comforted or rewarded with food, she will eventually start comfort eating of her own accord. The only difference between her and a child with a genetically determined large appetite is that the latter already has a desire to eat the chocolate and does not require any behavioural reinforcement to enjoy sweet and/or fatty foods. A child with a larger appetite who is additionally comforted with food will descend ever deeper into the downward spiral of obesity. So it's best to take countermeasures early.

But even if such measures were never taken in our childhoods, we are not lost causes as adults. Studies have shown that food preferences can still be influenced in adulthood and are not an inescapable fate. In one study, the brains of both obese subjects and subjects of normal weight were scanned to record their reactions to food. The reward centres in the brains of the obese subjects showed a strong reaction to food that is high in fat (fast food, sweets). The test was repeated after the subjects had followed a dietary plan containing healthy, low-calorie foods for several months. The reward centres in the obese subjects' brains reacted more strongly to healthy and low-calorie foods in the second test.

Alcohol addiction serves as a good example to explain the way people's genes can affect them. Everyone understands that there is no such thing as a genetic alcoholic — that is, someone who is an alcoholic without ever having drunk alcohol to excess. No one can be an alcoholic without ever having touched a drop. Despite this, alcoholism can be said to be about 50 per cent genetically determined. Herz (1997) writes that part of this genetic determination is to do with the various ways dopamine can be regulated in the brain and in certain enzymes in the body, which affect the way alcohol is processed. What this ultimately means is that alcohol has a considerably stronger effect on some people than others, and those people therefore have a greater risk of becoming addicted. A genetic inclination towards being overweight is similar: genes do not have direct influence, they don't *make* people fat or addicted to alcohol, they just create the conditions which influence risky behaviour.

New research is now emerging that could change our understanding of genes completely. It indicates that our lifestyle can activate or deactivate certain genes, and even that a mother's lifestyle can influence whether she passes on certain genes to her children or not (Watters, 2006).

In the end, our genes just set out the path we will follow if we don't actively strive to change its direction (and the change of direction, in turn, might influence the effects our genes have on us). Unfortunately, nature and nurture often go hand in hand since parents have the same genetic makeup as their offspring. Human beings also tend to seek out environments which correspond to their natural inclinations. This means that working against our genetic predispositions requires conscious effort, a kind of swimming against the tide, so to speak. However, studies have shown that those efforts are only temporary, because once we have become habituated to new behaviours, we no longer have to struggle to maintain them.

Gut bacteria make you fat

There was a story that recently did the rounds in the media of the case of a slightly overweight woman who received a bacteria culture transplant from her obese daughter and put on 16 kg in the following year. Gut bacteria were thought to be the culprits for her weight gain. Likewise, there are increasing reports about the possibility of using gut bacteria from thin people to treat those who are overweight.

These types of stories are often one-sided and over-simplified. They work on the assumption that differences in gut flora are the only factor responsible for a person's weight — just like with genes. But gut bacteria are once again only the moderators, which work indirectly by digesting certain foodstuffs better than others and thereby increasing their host's appetite for those foods. But this is not a one-way street; our eating habits also influence our gut bacteria, and they adapt to what we feed them in the long term.

In research carried out with beagles, for example, one group of dogs received unlimited food over a period of six months until they had gained an additional 67 per cent of their original weight. In the same six-month period, seven other beagles were fed normal amounts to keep them lean. By the end of the experiment, the guts of the obese dogs resembled those of obese humans: the bacteria were less diverse, and their distribution was altered (Park et al., 2015). The interesting point here is that the dogs all were given the same food, and only the amount they received was varied.

In 2015, *Quartz* reported a similar study using rats, which were fed either a balanced diet or a very high-fat one. The rats were allowed to eat as much as they wanted, and after a period of two months those on the fatty diet were around 25 per cent heavier than their peers in the other group. Just as in the case of the beagles, their gut flora had altered, and their vagus nerve, which is responsible, among other things, for transmitting signals that we are sated, was impaired, leading the rats to carry on eating even when they were full. The research group that conducted this study believes that the results can be transferred to humans and that an unbalanced diet can lead to a disequilibrium in bacterial populations, which in turn affects the signals for hunger and satiety sent to the brain, leading to more of the unbalanced food being eaten.

What this means is that when we alter our eating habits, we start out having to battle against our 'old' stomach and gut bacteria, but both adapt to our new diet in time. In the case of the woman described above who received the gut bacteria transplant from her daughter, I can imagine that the change in her gut flora increased her appetite for fatty and sugary foods, and following her appetite, she ate a more calorie-rich diet. However, if she had paid conscious attention to what she was eating, her gut bacteria would not have *made her fat*, but instead would have adapted to her lower-calorie diet over time.

[Fat logic for women] Childbirth automatically makes you fat

Pregnant women are eating for two. That being said, the unborn baby is not a two-metre-tall construction worker; it is so tiny that even at the end of a pregnancy, when the baby's need for food is greatest, the mother's body requires only 15 to 25 per cent more calories than usual. On average, pregnant women require an extra 225 kcal per day.

According to the guidelines for weight gain during pregnancy issued by the National Academy of Medicine, the recommended increase in body weight for underweight women is around 12 to 18 kg, for women of normal weight, the figure is around 11 to 16 kg, and for obese mothers-to-be, it is only 5 to 9 kg. Most of that is not the mother's fatty tissue, however, but rather the weight of the baby, the placenta, amniotic fluid and pregnancy-related water retention. Below is an approximate breakdown of the sources of pregnancy-related weight gain (as presented on www.medezinfo.de):

Mother	Child
womb, 970 g	foetus, 3400 g
breasts, 405 g	placenta, 650 g
blood, 1250 g	amniotic fluid, 800 g
water, 1680 g	

So on average, more than 9 kg of pregnancy-related weight gain is due to factors other than body fat. This means that overweight and obese women would likely actually lose some body fat if they remained within the recommended parameters for weight gain during pregnancy.

It makes especial sense for a woman's body to build up small fat reserves during pregnancy, as breast feeding is such a calorie-intensive activity: milk production can easily use up the same number of calories as exercising intensively for one hour a day. It's why most women lose weight while they're nursing, as long as they eat normally.

The main reason why many women gain body-fat weight due to childbirth is that pregnancy and breastfeeding cause hormonal changes within a woman's body that can strongly affect her appetite. As explained earlier in this book, an altered appetite can have a powerful effect on portion-size perception, creating the impression that the body's rate of metabolism has changed. This explains the claims you hear so often, along the lines of, 'I used to be able to eat whatever I wanted, but since I've had kids I really have to watch what I'm eating.' In fact, the body's metabolism will not have changed, or the possible changes are so slight that they account for only a few calories or even less — depending on the amount of weight (more weight → higher calorie consumption), muscle (possible less physical activity during pregnancy → muscle-mass reduction → lower consumption), etc.

Moreover, the changes in lifestyle and the increased stress that follow a pregnancy can bring with them the danger of slipping into new routines that could lead to weight gain (or loss).

But here, once again, the same thing is true: changes in body weight don't happen *on their own*, and pregnancy doesn't *make* women fat. If it does anything, it creates the conditions for a woman to eat more calories than her body needs.

You gain weight as you get older because your metabolism slows down

This is like saying, 'You go into debt as you get older because your pension is lower than your previous wages.'

As long as the amount of fuel you ingest is in line with the energy your body uses up, you don't put on weight. When you start finding yourself gaining weight, it's time to adjust either the number of calories you take in or the amount of energy your body uses, to regain the balance and put a stop to the weight gain. This is true for people in any life situation, whether the imbalance is due to a career change into a less physical job, a busier job with less time for physical exercise, or any other factor that alters the body's energy needs.

It's also a myth that increased age must necessarily cause the body's metabolism to slow down. It is true that, statistically speaking, it happens, which explains why the calculated bodily energy requirements produced by some websites and apps go down as the entered age goes up. But the reason for this is not connected to any biological condition, it's just the result of statistical observations related to the fact that, on average, most people become less active when they get older, their muscle mass is reduced, and their general lifestyle means they have a reduced energy requirement. As early as 1977, Tzankoff and Norris were able to confirm the theory that age-related drops in the body's energy needs can be explained almost completely by the loss of muscle mass. And in 1991 Vaughan et al. once again confirmed those results with a study that compared the energy expenditures of older and younger test subjects. On average, the older subjects had less muscle mass, as well as a higher proportion of body fat and lower energy consumption. However, when seen in proportion to their body mass, their energy consumption was no lower than that of the younger subjects. And so, once more, the deciding factor was not age, but physical activity and muscle mass.

Therefore, if you build up muscle mass or maintain existing muscles with weight training later in life, there is no reason why you should necessarily require less energy than you did when you were younger. But it should be said that it becomes increasingly difficult to develop muscle mass as we get older, so it's a good idea to build up *reserves* early, so that they can be maintained with increasing age.

My metabolism is slower than other people's

Yes, it is unfair that a small woman can't eat as much as a 190-cm-tall body builder. It's unfair, too, that I can't spend as much money as a top bank manager. Of course, I could insist on spending as much, but then I'd go into debt.

Metabolism is not like money, of which some people really do have too little to survive. In the case of metabolism, even the smallest, daintiest person burns well over 1000 kcal a day — and a lot can be done with that. Alternatively, you can use weight training to build up muscle mass and so increase the basic amount of energy the body needs to function. Endurance sports also increase the body's base energy needs, making the occasional 'extra' bar of chocolate a harmless treat.

I can't lose weight because I can't exercise

There are lots of different reasons why some people can't exercise or can only exercise in a limited way. But those reasons are irrelevant because exercising isn't necessary in order to lose weight. Fitness enthusiasts like to say things like 'You can't outrun a bad diet', and 'Sixpacks are made in the kitchen, not in the gym'. In general, the approximate rule is that diet is responsible for 80 per cent of body weight, and physical exercise is responsible for 20 per cent.

A quite recent study showed that many people who start exercising will even gain weight — because of an increase not in muscle mass, but body fat (Sawyer et al., 2014). This is because people tend to overestimate massively the amount of energy they consume while exercising, and then often eat far more calories than they would otherwise have done, reasoning that they have 'earned' them through sport. In addition, exercise increases the appetite. In a study by Finlayson et al. (2011), 41 per cent of participants in an exercise program had such a large appetite after each training session that they saw almost no improvement in their body weight. King and Blundell (1995) were also able to show in a study that their subjects had a stronger desire for fatty foods after a fifty-minute endurance training session than the subjects in the non-exercising control group. Although they had burned energy in the short term during the sports session, they overcompensated with a higher calorie intake afterwards.

In the first six months of my weight-loss process I did no exercise — apart from 20 minutes of gentle training with a rubber resistance band to stop my muscles from wasting away completely due to inactivity. Those twenty minutes of 'training' burned far less energy than an average inactive person burns in their daily lives, as long as they're not stuck on the couch. I still managed to lose around 45 kg in that time. My caloric deficit was large, of course (about 2500 kcal), but even with a moderate deficit of 1300 kcal per day, I would have lost more than 20 kg without moving a muscle.

In many cases, it's even advisable to wait till some weight has been lost before beginning an exercise program, as sport puts a great deal of strain on very overweight people's joints and cardiovascular system. Swimming is the exception to this rule, since the buoyancy provided by the water takes the strain off overburdened joints. But for very overweight people, going swimming can involve a huge emotional effort to overcome the hurdle of appearing in public in swimwear. For me, to be honest, swimming was never my sport. Luckily, cycling and walking are also relatively easy on the joints. Being overweight can also be a problem for cycling enthusiasts, though, because it's hard to find a bike that can carry 110 kilos or more. Reinforced bikes are often expensive. I eventually cobbled together my own recumbent home bike by placing an armchair behind an old exercise bike. At that time, the saddle would have been strong enough to hold me, but the lying-down version was much more comfortable and ideal for watching television while exercising.

But basically, for extremely overweight people, normal, everyday movement such as gentle fitness walking, or even just going for a stroll, is enough. Even a twenty-minute walk every day — whatever a person's weight — will have a hugely positive effect on the cardiovascular system. It's also helpful in combating stress and depression.

Exercise is generally a wonderful thing when it comes to health, fitness, and wellbeing, and it is absolutely to be recommended. But that's true for everyone, whatever their weight. Although exercising is extremely hard work for people with a certain body weight, there are certainly overweight people who enjoy sport and are passionate about it.

For an overweight beginner, though, sport is more likely to be a torment, and these people run a greater risk of injury or strain. Incidentally, overweight people don't find exercising torturous just because it is harder work moving so much bulk. Recent studies have found that hormones formed in fatty tissue can influence the chemistry of the brain in such a way as to suppress the so-called 'runner's high' — the feeling of euphoria that can follow after endurance sport (Fernandes et al., 2015). This means that there are two hurdles for overweight people to overcome when they start to do sports, since they don't even get the reward of the happy hormones that can help us to push past difficult moments during exercise. The silver lining, of course, is that overweight people who take no pleasure in exercise can come to love sport as soon as they've lost some weight.

The basic rule is that there is no need to force yourself to exercise if it really is pure torture and no pleasure at all. There is one exception to this rule: weight training makes absolute sense for people of any body size, in my opinion. There are several reasons for this:

The muscles that are built up and/or maintained through weight training protect the joints from strain, which is particularly important for overweight people.

Muscle mass can counterbalance the detrimental effects of fatty tissue, for example, by reducing inflammation processes.

As muscle mass increases, it becomes easier to shift the body's weight, making other types of exercise easier. And movement becomes more pleasurable in general, increasing the motivation to engage in other types of physical activity.

Muscle mass increases the body's basic energy needs and so helps to stop those energy needs from falling too much due to weight loss (or, if no weight loss occurs, it even increases energy needs).

Anyone can do weight training, no matter how heavy they are, how fit they are, or whether they have injuries or other factors limiting their ability to exercise — there will always be a 'work-around' for any situation. Even people who are bedridden can help strengthen their bodies with regular, small weight training exercises.

Weight training doesn't have to be in a gym, nor does it require professional equipment: there are many sources on the internet or in libraries with advice on how to do simple but effective exercises in your living-room at home. That being said, a gym is not a bad place to start, as the training machines guide your movements. And even in inexpensive gyms you can find trainers available to give advice and tips. For my part, for what it's worth, I have never experienced negative treatment at the gym. The opposite, in fact. Even when I weighed 140 kg, everyone there was always friendly, and most people were happy to do their own thing and pay no attention to others. In my experience, the fitness community is very open to newcomers and always willing to help or advise. Most people are happy when other people take an interest in their passion.

The bottom line is that exercise isn't necessary to lose weight, but even with relatively little effort, it's possible to improve your fitness levels and profit from an increase in your muscle mass.

BMI is bullshit: my weight is muscle

I think one of the biggest lies I used to tell myself was that although I was fat, at least I was (also) strong. My first few sessions at the gym were bad. Not because anyone was mean to me, but because I was forced to admit that I wasn't particularly strong after all. True, I was able to leg press a relatively large amount (a little over 100 kg) — not bad for a woman who'd never done any training. But when you consider that it wasn't even three-quarters of my own body weight, it actually wasn't very much at all. For comparison, I am now able to press 130 kg — about twice my own body weight.

As an alternative to the BMI, there is now a newly developed method of determining how fit you are (as a fat person). It's called the sitting-rising test (SRT). It involves standing barefoot and in comfortable clothing in a clear space, then sitting down and standing up again in one motion without using your hands to support yourself (by touching the floor or other parts of your body). You begin with a score of ten points, and a point is deducted for every time you use your hands for support, while half a point is taken off for every wobble due to loss of balance. A score of less than eight indicates a doubled risk of dying in the next few years; a score of less than two points indicates a ten-fold increase in the risk of dying soon.

When I still weighed 150 kg, I usually found it difficult to get up from a flat floor if there was nothing I could use to haul myself up with. For a long time, I couldn't even imagine how I would get up without using my hands to support me. To be honest, I had to watch some videos online to learn the correct way of executing this test before I was eventually able to score ten points — by which time I weighed 65 kg.

As for criticism of BMI measurements: in my opinion it's directed in the wrong place. It's true that the Body Mass Index does not differentiate between fat and muscle mass, as it just expresses the ratio between body weight and height. But the BMI was devised in a time when people were considerably more active than they are today. A 2012 study by Shah and Braverman combined BMI with body-fat measurement and found that the BMI is actually extremely inaccurate — but only when it comes to diagnosing excessive fat. Of the test subjects whose BMI indicated that they were obese, only 1 per cent had a normal amount of body fat. This means that only 1 per cent of subjects were muscular enough to render their BMI wrong. When divided by sex, this figure rose to 3 per cent among males, and sank to zero among female subjects. What this means is that when a BMI figure predicts that a person is obese, there is a 99 per cent chance that they really are obese. And, among the 1 per cent whose BMI is skewed by their muscle mass, those muscles are clearly visible for everyone to see.

Far more interesting, however, is the fact that the BMI is actually wrong in two cases out of every five — when it predicts that a person is *not* overweight. In fact, 39 per cent of the subjects in the abovementioned study had excess body fat even though according to their BMI they were not overweight.

An analysis of several studies of the BMI (Okorodudu et al., 2010) came up with a similar result: it also showed that the BMI was an accurate predictor of adiposity (obesity) 97 per cent of the time that it showed someone to be overweight, but the analysis also found that it assigned normal weight to about half the people who actually had too much body fat.

A study carried out by Finnish researchers (Männistö et al., 2014) found that 34 per cent of male subjects and 45 per cent of females were obese according to body-fat measurements, despite having a healthy BMI. As it happens, there is a problem with BMI–body-fat comparisons in that body-fat charts usually only include the categories 'normal' and 'obese', with slightly overweight people being placed in the normal group. So the proportion of those who have 'slightly increased body-fat levels' despite their normal BMI may often be even higher.

A German study by Hauner et al. (2008) investigated the prevalence of obesity as measured by BMI compared to waist circumference and came to a similarly drastic conclusion: while 'only' about a quarter of the subjects were obese according to their BMI, some 40 per cent were classified as obese on the basis of their waist measurements. Women were more likely to be obese according to their waist circumference, men were more likely to be obese according to their BMI. Abdominal girth is a good method of determining body fat, insofar as it accounts for the fact that very fit people, whose weight is increased by their high proportion of muscle mass, tend to have small bellies. So the results of this study once again show that BMI measurements are more likely to *under*-represent the number of overweight and obese people in our society, rather than judge muscular people too severely.

To add to this, in their 2015 study, Pellegrinelli et al. found that fat causes inflammation and atrophy in muscle cells. In effect, this means that the presence of fat leads to loss of muscle mass. So, even though overall muscle mass will grow along with the rest of the body because of the increased strain put on it as body mass increases, that growth is then partly negated by the damage actively caused by fat. In relative terms, that means increasing body mass tends to result in decreasing muscle strength. This effect is even more pronounced in people with low body weight but excessive fat.

Unless we are talking about body builders, the rule is that if your BMI places you in the overweight or even obese category, we can be pretty sure that you have too much fatty tissue in your body. And a BMI within the healthy range can still mean a person is carrying too much fat on their haunches.

Being overweight isn't even that bad for you

This is the fat-logic argument I encounter most often, and which I believed (*wanted* to believe) myself for many years. It is also the one I kick myself about the most, in retrospect. I always claimed to have made a rational decision about my weight ('I enjoy eating too much and weighing less is just not worth the inevitable restrictions it would involve.') But I was labouring under two misapprehensions: that it is extremely difficult to achieve and maintain normal weight; and that it doesn't have all that many advantages anyway. Now, I argue the opposite whenever I can.

I respect anyone's decision to set other priorities and happily accept being overweight or obese. Just because you *can* change a situation, it doesn't mean you *must*. With that said, I do think it's important for that decision to be an informed one, rather than one based on self-delusion or distortions of the truth by others. If it isn't, you run the risk, later in life, of thinking, *If I'd properly examined the facts, I would have chosen differently.*

It is for this reason that I present, in the chapters that follow, the results of my research into the consequences of obesity. I've often received feedback saying that the medical section of my blog is a difficult read, especially the part covering individual medical conditions, and I'm sure this is true. But I have always considered it, and still consider it, important to deal with this rather dry subject in as much detail as possible in order to really drive home the point that this is not about whether your bum looks better as a size 36 or a size 42. Rather, it's about what goes on inside our bodies, and about how being overweight directly affects our quality of life. For this reason, I recommend that you at least skim through the following sections, because, although you may be thinking *Yeah, yeah, I know all that*, you might still find something out that you didn't know before. At least, that was my experience while I was researching this part of the book.

But my blood-test results are great!

Obesity is a bit like smoking: the tumours don't start growing right after the first cigarette. For someone who is naturally prone to lung problems, it might take five years. Another person's lungs might be able to take fifty years of constant damage. But just because the damage isn't visible, it doesn't mean that it isn't there. A relatively recent analysis from 2013 investigated the long-term consequences of obesity with the specific aim of examining so-called healthily obese people. A comparison between healthy people of normal weight and healthy but obese subjects showed that the latter group had a significantly higher risk of dying or developing cardiovascular disease. The scientists who carried out the study therefore came to the conclusion that the belief that you can be 'fat but fit' is just a myth (Kramer et al.).

A still more recent study by Bell et al. (2015) confirmed those results. It followed supposedly healthily obese subjects over 20 years and found that more than half became *un*healthily obese in the course of that time. Their risk of becoming ill was eight times higher than that of the healthy group with normal weight.

A Korean study covering 15,000 test subjects came to a similar conclusion: obese people with good blood-test results were found to be more likely to suffer from coronary artery calcification, leading to a higher risk of cardiovascular disease (Chang et al., 2014).

A study by Appleton et al. (2013) also showed that 'healthily obese' people had twice the risk of developing bad blood-test results or diabetes within five years when compared to people of normal weight.

In this context it should also be noted that these studies compared the healthiest obese people with healthy people of normal weight, meaning there was already a degree of preselection. Overweight and obese people more generally also includes people with risk factors such as high blood pressure or increased cholesterol levels. These tests only show that obese people with blood-test results within the normal range still have a higher risk of illness or death due to their extra body fat. As for the rest — they're even worse off, health-wise.

What about overweight athletes?

Whenever the conversation turns to the health risks of being overweight, the example always comes up of professional athletes who are fit despite their large body mass — like sumo wrestlers and American football players. Athletes are seen as the epitome of fitness, and it's true that professional sportsmen and women do have a higher average life expectancy and a lower risk of cardiovascular disease than others. According to a study by Sarna et al. (1993), on average, endurance athletes live five and a half years longer than non-athletes, those who play team sports live four years longer, and weightlifters can expect to live around one and a half years longer than the norm. After analysing several studies, Clarke et al. (2012) reached a similar conclusion, finding that Olympic athletes live on average almost three years longer than others.

So the argument goes that, if there can be severely overweight professional athletes, then being overweight cannot automatically be unhealthy, right? Wrong, unfortunately. The higher life-expectancy figures don't apply to obese athletes. According to the US Centers for Disease Control and Prevention, for example, American footballers with a BMI of over 30 were twice as likely to die from a heart attack, and the particularly heavy linemen died on average 25 years earlier than the average population. Furthermore, a study of ex-linemen in their fifties carried out by Hurst et al. (2010) found they had the same highly increased risk of cardiovascular disease and metabolic syndrome as other obese people.

Sumo wrestlers also have a significantly reduced life expectancy compared to the normal population, and the higher their BMI is, the more their morbidity increases (Hoshi & Inaba, 1995).

This clearly shows that not even the biggest protecting factor — physical exercise — which makes for a significant increase in the life expectancy of other athletes, is able to compensate for the harmful effects of increased body fat. Despite their relative fitness, the bodies of obese athletes are placed under enormous strain because of their sheer mass, and that is especially true for their cardiovascular system.

It's actually healthier to be slightly overweight than to be normal weight

This myth is kept alive by the fact that, every so often, a study is published in which the mortality rate of 'slightly overweight' people (BMI 25 to 29.9) is found to be lower than that of people of normal weight (BMI 18.5 to 24.9). Flegal et al. (2013) is one of these studies. But their research failed to take two important factors into account: firstly, smokers weigh less on average than non-smokers, which meant that the group of people of normal weight included more smokers (who had a higher risk of death due to their smoking habit); secondly, a lot of people lose a significant amount of weight in the course of a severe illness and die with what appears to be normal body weight, even though they were overweight when they first became ill.

My grandmother is an example of this: she was severely obese, suffered a stroke, was bedridden for years, and lost a lot of weight, finally dying with her weight within the normal range. It is the conditions that are associated with obesity in particular, such as cardiovascular disease, cancer, and strokes, that often lead to severe weight loss prior to death. When these people, i.e., those who weigh less due to illness or because of their nicotine consumption, appear in greater numbers in the group of people with normal weight, the morbidity results will, of course, be skewed.

If these kinds of factors are controlled, the results look very different. Once that's done, being even slightly overweight no longer appears to be healthier than being of normal weight, but rather more harmful. Not harmful to the same extent as actual obesity, of course, but it's still unhealthy. In 2006, Adams et al. investigated whether or not being 'slightly overweight' is harmful to health, and took into account factors such as chronic illness and smoking. According to their analysis, the subjects' risk of dying within the subsequent ten years after the start of the study was increased by 20 to 40 per cent among those who were slightly overweight, while for the obese it was increased by 200 to 300 per cent. It's convincing evidence of the fact that being even slightly overweight constitutes a health risk.

A 2010 study of almost 1.5 million people carried out by Berrington de Gonzalez et al. also found that people with a BMI of between 20 and 24.9 have the lowest risk of mortality. In the underweight range (BMI 15 to 18.4) the risk of death was increased by 47 per cent, placing it at a similar level to those who are slightly overweight (BMI 30 to 34.9). Severe obesity was associated with a 151 per cent higher risk of death.

Calle et al. (1999) found that the BMI with the lowest risk of death among healthy female non-smokers was between 20.5 and 24.9, and for equivalent men, it was between 22 and 26.4.

A study by Manson et al. (1995) found that for women who had never smoked and weren't underweight due to illness, the lowest risk of death was with a BMI of less than 19. A BMI of 19 to 25 meant a 20 per cent increase in the risk of mortality, compared to the average across all subjects, a BMI of 25 to 27 increased the risk by 30 per cent, and the increased risk for a BMI between 27 and 29 was 60 per cent. Finally, among obese women, the risk of death was increased by 120 per cent. The risk of dying from cardiovascular disease was as much as 310 per cent higher for obese women than for those with a BMI of less than 19.

In a 2012 interview with the German weekly newspaper, *Die Zeit*, the epidemiologist Dr Rudolf Kaaks said that even a BMI in the upper-normal range can represent a far higher risk of getting cancer than one in the lower-normal range. This is presumably linked to the previously mentioned problem with the BMI, in that half the time it assigns harmless weight levels to people whose percentage of body fat is in fact too high.

In my opinion, body-fat percentage is ultimately a more crucial factor than BMI. In large-scale studies like those quoted above, and in the studies that I quote later in the sections about individual medical conditions, a BMI at the lower end of the normal range almost always turns out to be associated with the lowest risk. This is probably thanks to the fact that the majority of people in modern societies lead very physically inactive lives.

In 2015, the British newspaper *The Telegraph* ran an article saying that previous recommendations of 2.5 hours of exercise per week were *unrealistic* and *too challenging*, and suggesting that 20 minutes per week was a better target. In a survey on the *Citimarathon* blog, 60 per cent of Germans said that they did no exercise, and only 20 per cent actually exercised regularly. This must lead us to understand that the majority of the population has very little muscle mass, increasing the probability that they have too much body fat and an increased BMI — even if it is still within the normal range.

As I already mentioned in the context of the BMI, those people categorised as overweight due to their excessive muscle mass are the absolute exception, while far too many people end up in the normal weight category simply because they have so *little* muscle mass. So, an average person who leads a relatively sedentary life and does little or no exercise would need to be at the lower end of the BMI scale to have the best chances of a good body-fat percentage and low risk to health. Someone who exercises and is reasonably muscular, by contrast, can have a healthy body-fat percentage while in the mid to upper normal BMI range, or, for those who do a lot of weightlifting, even in the slightly overweight range.

In general, the situation is very clear: a low percentage of body fat is the healthiest state, and for anyone who doesn't happen to be a body builder, normal weight is considerably healthier than being even slightly overweight.

Just because excess weight is associated with illnesses, that doesn't necessarily mean it causes them

In fact, there did appear to be evidence that certain genes associated with obesity could also increase the risk of developing diabetes — and recent research that took into account actual BMIs then showed that this connection only develops in people who are overweight. Carriers of the gene who were within the normal weight range were found to have no higher risk of diabetes than non-carriers (Sandholt et al., 2012). As already explained in the chapter on genetics, genes just increase or decrease the probability that we will create certain conditions through our *behaviour*, but that behaviour is ultimately under our own control. We are not controlled by our genes like a marionette on the strings of a puppetmaster.

There are now a wide range of studies available that show that people suffering from conditions associated with obesity can achieve great improvement, or even complete recovery, by losing weight. Dhabuwala et al. (2000) found that severely obese patients who slimmed down to a BMI slightly above normal saw large improvements in their hypertension within two years. Of the 42 patients receiving drugs to treat high blood pressure, 18 were able to come off their medication completely. And of the 34 asthmatics, half were able to stop their medication. Eleven patients suffering from sleep apnoea found themselves completely cured of the condition. I will cite more studies showing improvements in various illnesses from weight loss in the section where I cover individual medical issues.

Overall, the research in this field is clear. The fact that conditions associated with obesity can be improved or cured by losing weight shows that excess weight is indeed the decisive cause of the condition, rather than bad luck or genetic predisposition.

There is also evidence to disprove the argument that losing weight might not be the reason for the improvements reported, and that it might rather be the fact that the patient is eating more healthily or doing more exercise. Wood et al. (1988) studied a group of men for more than a year as they lost weight using different methods. Some used caloric reduction, while changing the nature of their diets as little as possible. Another group were asked to continue eating as much they previously had, but to begin exercising. The two groups lost similar amounts of weight and saw similar improvements in their blood-test results. The men who lost weight by exercising did not see any better results than those men who had reduced their food intake, but, apart from the loss of weight, whose lifestyle had not changed.

Thin people can get sick, too

This habit of downplaying dangers is the same for all behaviours that are named as unhealthy, such as smoking, drinking, and taking drugs — and, as I have already mentioned, I convinced myself it was true for long enough. Having one or even several risk factors, of course, doesn't mean that you will necessarily be affected, but the fact is that your risk is increased — especially when it is already higher for other reasons.

Oh man, I was just at the doctor's, and he lectured me as if I was about to drop dead on the spot, just because I smoke and I'm overweight. Typical doctor's nonsense. I mean, come on, my grandad lived till he was 89, and he was a smoker, *and* he had a big belly …

… And I had a neighbour who was one of those uber-healthy ecofreaks who ran marathons and everything. Then he got cancer at the age of 40, and six months later he was dead. So much for a 'healthy lifestyle'. It's a load of bullshit.

Yeah, totally. My boss insists that we take safety precautions when we're working high up, but my aunty died in the shower. And recently someone fell from the sixth floor and didn't hurt himself at all. So much for 'death by falling from a great height'.

Different people are built differently and have different weak spots. I, for example, am prone to urinary tract infections, apparently because my urinary tract has an unfortunate kink in it. Sitting on cold surfaces increases the risk of contracting urinary infections for anybody, but for me, five minutes is enough to give me an infection, where others might only start to find it a problem after five hours. Of course, someone could easily come along and say, 'I often sit on the cold floor, but I never get urinary tract infections, and Nadja hardly ever sits on cold surfaces — and when she does, it's only for five minutes — and she's always getting urinary tract infections. Clearly, cold surfaces have nothing to do with it. In fact, it might be better for Nadja if she sat on the cold floor more often.'

Anyone who is prone to certain health problems — if their parents or grandparents suffered from diabetes or heart disease perhaps — or anyone who has sustained an injury that made them liable to problems with their joints, will be all the quicker to feel the consequences of the permanent strain that excess weight puts on their system.

A study by Grover et al. from 2015 compared the health and life expectancy of people in different weight categories. The interesting thing about this study is that it counted not only the years of life lost, but also the number of healthy years lost, thus taking into account quality of life to some extent. The results concern only the greatest threats to life, namely diabetes and cardiovascular disease. The study found that slightly overweight people, if they were already overweight as young adults, lose up to 2.7 years of life and 6.3 healthy years. For obese people, the loss of life-years rises to up to 5.9 and the loss of healthy years can be up to 14.6. For severely obese people, death comes up to 8.4 years earlier, and an average of 19.1 years of health are lost.

If it was just a question of when you meet your death, some might say, 'Okay, so I'll die five years earlier, but at least I'll have enjoyed life!' But when you spend the final decades of your life bedridden and in chronic pain, which often involves being permanently reliant on care, there cannot really be any talk of enjoying life.

Personally, I hate having to ask other people for help and I can't stand not being able to do things for myself, so I must admit that I find the loss of healthy years to be a much more persuasive argument than 'just' the prospect of maybe dying a couple of years earlier. In my professional life, I have worked with patients suffering from chronic pain and illness and I know from first-hand experience how much that can impinge on those people's lives, especially as their condition worsens.

In my own case, by the way, the phrase 'Okay, so I'll die five years earlier,' was a huge understatement, since I already had a BMI of 50 when I weighed 150 kilos. According to a new study (Kitahara et al., 2014), being extremely overweight can shorten a person's life by up to 13.7 years, making it more deadly than smoking.

Who says it's definitely the weight that makes you sick?

Lots of people argue that it is not excess body weight itself which makes people get sick, but the fact that it often results from an unhealthy lifestyle. They certainly grant that being overweight comes with health risks, but they think that that only applies to 'stereotypical fat people' and not to people who are overweight but lead a physically active life and eat healthily. I used to believe that, too. I thought that because I never ate fast food, and because I avoided fatty meats, often cooked for myself, and ate a lot of fruit and vegetables, that I was one of those exceptions.

Many people overestimate how much you have to eat in order to get (so severely) overweight. In the past, I would put on an average of around 1 kg a month, which corresponds to about 200 kcal a day. Of course, with a weight of 150 kg, my body's requirements were already very high, at around 2700 kcal per day. That's the same amount as a tall man who has a physically demanding job or works out. In order to gain (or maintain) such a high weight, you don't have to be continuously stuffing yourself 24 hours a day. If you consider that a spoonful of oil alone contains over 100 kcal, or a pack of trail mix can have as many as 1000 kcal, it's enough just to have several smaller excesses throughout the day, which you might not even realise are excesses.

As it happens, I believe that this image of the eating

behaviour of (very) fat people is what leads many to see themselves as an exception to the rule and to say, 'There *must* be something wrong with me — I don't eat like a typical fat person!' In fact, hardly any fat people eat like a typical fat person.

Few people watching me eat my pumpkin seed roll with cheese (about 400 kcal) in the morning, my previously-described salad (around 1500 kcal) for lunch, and a pack of trail mix over the rest of the day (about 1000 kcal) would have said my eating behaviour was unusual. Lots of people would even have got the impression that my daily food intake was not that much, and was quite healthy, with those wholemeal rolls, salads, and nuts.

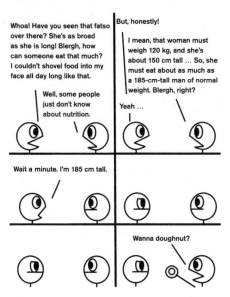

As soon as you start to understand how fatty tissue works and the effects it has on the body, your perception begins to get clearer.

Before I started researching this topic, I thought that fatty tissue was just an energy store that was 'attached' to the body, so to speak. I now know that fatty tissue is an organ that produces sex hormones — among other things. This explains why both underweight and overweight women can experience problems with their menstrual cycle, because their bodies are either producing too much or too little of certain hormones. So it *is* necessary for the body to maintain a certain amount of fat, not only as an energy reserve for leaner times, but also to keep the body functioning normally. At the same time, fatty tissue is made up of cells that die off and need to be replaced. Dying fat cells are recognised by the immune system as foreign bodies, and that can lead to inflammation reactions. These infections in turn have an effect on the rest of the body's systems and can cause conditions like diabetes and inflamed joints.

What's more, fatty tissue builds up both *on* and *in* the body's organs, such as the heart and the liver, and adversely affects their ability to function properly. The size of the affected organs is increased by the fatty deposits, pushing other organs aside. An enlarged heart, for example, can displace the lungs outwards, and, although the ribs are able to expand, the lungs aren't able to inflate as far as they can in a person of normal weight. On one hand, an overweight body requires more oxygen to supply all that extra tissue, but on the other hand, its lungs are unable to function to full capacity. This leads to an oxygen deficiency and less energy overall, as well as tiredness and exhaustion.

All this is compounded by the fact that overall body mass is higher, and so the organs have to work harder to supply it all with blood and nutrients. The heart not only becomes enlarged because of the fat deposits on it, but also has to work harder in order to pump more, which leads to more rapid wear and tear.

Physically, the increased weight also puts more pressure on the body's joints, and that also causes wear and tear. Combined with higher inflammation levels, that wear and tear can considerably increase the risk of joint problems.

Beyond a certain body mass, and depending on the distribution of fatty tissue in the body, fat can also push outwards — for example, when a person is lying down. When this happens, the fat puts pressure on the lungs and airways, making breathing in certain positions much more difficult.

The risk of skin problems is increased not only by defective hormone production, but also by the fact that there are more skin-on-skin contact points, or 'fat rolls', where sweat and bacteria can gather.

Below is a round-up of the main medical complaints associated with excess body weight.

Diabetes

The inflammations caused by (abdominal) fatty tissue destroy cells and reduce the body's ability to respond properly to insulin and therefore to process all carbohydrates in the blood. This results in permanently increased blood sugar levels, causing damage in other parts of the body — it destroys the nerves in the extremities, for example, and can also affect the optic nerves. In mid-2014, the medical specialist Dr Max Pemberton wrote that he would rather have AIDS than diabetes, since, contrary to the prevailing public opinion, HIV is now an easily treatable condition and, from a medical point of view, it no longer leads to a reduction in quality of life or life expectancy. By contrast, diabetes is often dismissed as an 'irritant' although it not only often leads to a massively increased risk of strokes and heart attacks, kidney damage, blindness, and leg amputations, but is also associated with a reduction in life expectancy of up to ten years.

In 2015, Menke et al. published a study that claimed that half the US population already has at least prediabetes, and 12 to 14 per cent have the full-blown disease. The figure for Germany is not that high at the moment; according to Stock et al. (2005), it stands at 6.5 per cent. However, research carried out by Rathmann et al. (2003) indicated that among the fifty-five- to seventy-four-year-olds examined, 40 per cent had increased blood sugar levels or diabetes, half of which cases were undiagnosed. So, we must assume that the actual numbers are much higher. In the study by Menke et al., the results were also broken down according to BMI, revealing that the prevalence of diabetes among the subjects of normal weight was around one in 20, while of the obese subjects, one in five had the disease.

One long-term study (Colditz et al., 1995) followed more than 100,000 women over the course of 14 years. It revealed that BMI was the main risk factor in the development of diabetes. The risk of contracting the condition rose along with BMI, and even women in the upper-normal range (BMI around 24) ran an increased risk of developing diabetes.

Compared to women with a stable body weight, the risk of getting diabetes was increased by 90 per cent in women who gained 5 to 8 kg. For women who put on 8 to 11 kg, the risk of diabetes rose by 270 per cent. Losing more than 5 kg, on the other hand, halved the women's risk.

An analysis (Guh et al., 2009) of 89 studies showed that obese men had an approximately six times greater risk of diabetes than men within the normal weight range. The risk for obese women was no less than 12 times higher.

But this is not just a problem for adults: although type 2 diabetes used to be known as 'adult-onset diabetes', many children and adolescents are now affected by it, thanks to increasing rates of obesity among those age groups. According to research carried out by the Pediatric Academic Societies, the rate of diabetes among children rose by 30 per cent between 2001 and 2009. Almost one child in every 2000 has type 2 diabetes. American media recently reported the case of an obese, three-and-a-half-year-old child who was diagnosed with type 2 diabetes (*Medical Daily*, 2015).

There are now plenty of studies showing that type 2 diabetes can be prevented, or its risks highly reduced, by losing weight. For example, in a study lasting almost three years and published in 2009, one group of researchers compared the effects of an intensive change in lifestyle, including weight loss, with treatment with medication and with a placebo. Compared to the placebo, the medication resulted in a 31 per cent reduction in the risk of diabetes, while the lifestyle change lowered the risk by no less than 58 per cent. Even ten years later, the lifestyle-changing group showed the best results of the three, despite regaining some of the weight they had lost.

Clearly, even losing 10 per cent of your body weight has an extremely positive effect on the treatment of diabetes and other conditions and results in an increased life expectancy (Goldstein, 1992).

The US Diabetes Prevention Program (2009) recommends weight loss of 5 to 7 per cent of body weight and 150 minutes per week of moderate exercise, such as walking, to prevent diabetes.

Cardiovascular disease

If the heart has to supply a larger body mass, it has to increase in size itself in order to be able to provide that extra power. Enlargement of the chambers of the heart impairs the organ's ability to contract properly. Over time, this can lead to a situation where blood isn't completely pumped out of the chambers of the heart, which means that there's always some residue. The greater amount of blood being pumped through the enlarged heart leads to an increase in the pressure in the blood vessels, which in turn leads to hypertension.

At the same time, fatty tissue can damage the kidneys, which would also normally help to regulate blood pressure. Hypertension doesn't usually result in any symptoms itself, but in the longer term, it can cause damage to the cardiovascular system.

In a recent study (Pesheva, 2014), the blood of 9500 healthy men and women was tested for the enzyme known as troponin T, which is released by injured heart-muscle cells. Levels of the hormone were found to increase proportionally with increasing BMI. Over the next 12 years, nearly 10 per cent of the test subjects developed heart problems. That risk was almost double for the severely obese with a BMI of more than 35. The risk rose incrementally with BMI, growing by 32 per cent for every five-unit increase in BMI.

In an interview with *HealthDay* (2013), Reis and colleagues reported on a 30-year study which began in the 1980s. This research also focused on the risk factors for cardiovascular disease. At the start of the study, none of the 3300 participants were obese. By the end, around 40 per cent of the study subjects had become so. The results showed that each year of obesity was associated with about a 2 to 4 per cent higher risk of heart disease. The investigators found that 27 per cent of the long-term obese participants showed signs of heart disease.

In the analysis by Guh et al. (2009) mentioned above, the risk of developing cardiovascular disease was found to be 29 per cent higher for overweight men and 72 per cent higher for those men who were obese. The increased risk was even greater for women — 82 per cent higher for those who were overweight, and no less than 269 per cent higher for obese women.

Similar results were reached by another long-term study, this time in Sweden, which ran for 23 years and followed around 22,000 men, who were aged between 27 and 61 at the start of the project. It found that, in comparison to subjects of normal weight, the risk of suffering a heart attack was 24 per cent higher for the overweight men, and

76 per cent higher for those who were obese. If obesity was not compounded by other risk factors (diabetes, increased heart rate or blood pressure, smoking, high blood lipid levels, etc.) the risk of a heart attack was only slightly increased, but that was the case for only about 16 per cent of the obese men.

When it comes to risks for the cardiovascular system, experts also stress the benefits of a 5 per cent reduction in body weight, and for women in particular, a low BMI score is associated with the best protection against heart disease (Eckel & Kraus, 1998). This once again illustrates the fact that, for women especially, even a body weight within the upper-normal range can be associated with too much fatty tissue and increased health risks. Women with no previous history of heart disease were followed in a 14-year study. It showed that the group with the lowest risk of a heart attack consisted of women with a BMI of less than 21. Even with a BMI of only 21 to 22.9, the risk increased by 19 per cent. In the category of women with a BMI of 23 to 24.9, the risk was 46 per cent higher, and for the overweight women, with a BMI of 25 to 28.9, the risk was doubled. For those with a BMI of more than 29, the risk of heart attack was 350 per cent higher than the women with a BMI of 21.

Changes in body weight were also considered in the research. Similar to the diabetes studies, it was found that a stable body weight was associated with the lowest health risks. A weight gain of only 5 to 7.9 kg led to a 25 per cent higher risk of heart disease. A weight gain of 8 to 10.9 kg meant a risk increase of 64 per cent, and 11 to 19 kg of added body weight raised the risk by 92 per cent. Subjects who put on more than 20 kg had a 26 per cent higher risk of heart disease (Willet et al., 1995).

Fitness and regular exercise are an important protection factor, especially against cardiovascular disease. Research shows that a higher level of fitness correlates with a lower risk of heart attack — irrespective of BMI. However, obese men with the highest fitness level are in a higher risk group for heart disease than men of normal weight with the lowest fitness level (Högström et al., 2004). This shows clearly that a greater body mass puts more strain on the heart, even if the person concerned exercises frequently and has a lot of muscle mass. It's the increased body mass itself which causes the damage, irrespective of lifestyle.

I can say from personal experience that, although I was able to lower my very high blood pressure slightly by drinking green tea and going off the pill, it was only from losing weight, without exercising, that it reached normal levels. After losing 40 kg, my blood-pressure readings were no longer too high, and nowadays they're within the ideal range. My resting pulse rate used to be 90 beats per minute, and now, after exercising and reaching a normal weight, it's about 50 beats per minute. My heart no longer has to work so hard to supply my body with blood. The University of Göttingen's information brochure on blood pressure, published in 2005, says that for every kilogram of body weight lost, blood pressure drops by an average of 2 mmHg (millimetres of mercury, the normal unit for measuring blood pressure), and so 10 kg of weight loss are associated with a drop in blood pressure of around 20 mmHg.

CNN (2012) reported on a study of overweight subjects who had lost more than 5.5 kg over two years, and who not only prevented the increase of plaque deposits in their arteries but were even able to reduce them. Irrespective of the kind of diet they followed to lose weight, the subjects with the most weight loss showed the greatest subsequent reduction in blood pressure and improvement in arterial health.

Participants in a study by Merino et al. (2012) achieved equally significant improvements in arterial health and other cardiovascular risk factors in just three weeks, on a diet of fewer than 800 kcal a day.

Cancer

Shortly before I began writing this book, the media reported a new study by the WHO that attributed various kinds of cancer to obesity (Arnold et al., 2014). The authors said that overweight was the trigger for almost half a million new cases of cancer every year around the globe. This was especially the case for cancers of the uterus (62 per cent increase in risk for every 5 BMI points gained), gallbladder (31 per cent), kidneys (25 per cent), cervix (10 per cent), thyroid (9 per cent), and leukaemia (9 per cent). The Harvard University cancer specialist, Jennifer Ligibel estimates that overweight will soon be the leading cause of cancer around the world, surpassing even smoking. On an individual level, smoking is more carcinogenic, says Ligibel, but obesity is now so widespread that it causes more cancers overall.

Scientists are still researching the precise processes by which excess weight contributes to the development of cancer. One theory is that hormonal influences probably play a role, another is that it may be in the nature of fat tissue itself to promote the growth of tumours. Seo et al. (2015) discovered through animal experimentation that adipose breast tissue is denser and stiffer than other tissue, and that those characteristics promote the growth of breast tumours. Some of this effect was able to be reversed though weight loss.

Guh's analysis (2009) quoted the following cancer risk increases due to overweight:

	Women		Men	
Cancer	BMI 25–30	BMI 30+	BMI 25–30	BMI 30+
Breast	8 %	13 %		
Uterus	53 %	222 %		
Ovaries	18 %	28 %		
Colon	45 %	66 %	51 %	95 %
Oesophagus	15 %	20 %	13 %	21 %
Kidney	82 %	164 %	40 %	82 %
Pancreas	24 %	60 %	28 %	129 %
Prostate			14 %	5 %

What's more, a meta-analysis carried out by Niedermaier et al. (2015) indicated that the risk of developing a brain tumour also increases with increasing body weight. Overweight people contracted them 21 per cent more often than those of normal weight, and obesity increases the risk of a brain tumour by as much as 54 per cent.

Sleep apnoea

Sleep apnoea is a condition that is characterised by — mostly unnoticed — pauses in breathing of several seconds to several minutes during sleep, which can be extremely dangerous. Depending on the severity of the condition, it can lead to fatigue, drowsiness during the day, lack of concentration, and even heart problems. One of the causes is a weakening of the muscles of the respiratory tract, which can no longer provide the necessary resistance to keep the airways open and tend to 'give way'. Obesity can be a reason for this, if it places such a burden on those muscles that they can't push back. When it's not the cause, excess weight can still be a contributing factor, as it forces otherwise weakened muscles to work harder.

In a rather old study from 1985, losing 10 kg was shown to lead to a great improvement in the symptoms of 15 sleep apnoea patients (Smith et al., 1985). In 1991, Schwartz et al. confirmed these results, showing that weight loss yielded far better outcomes than other treatment methods.

More recent studies from 2014, e.g., Chirinos et al., have also shown the advantages of losing weight as compared to the administration of oxygen: they randomly assigned 181 test subjects to one of three groups: a weight-loss group, an oxygen-mask group, and a group which combined both treatment strategies. The weight-loss group saw the greatest improvement in symptoms like insulin resistance and poor blood-test results, irrespective of whether they were supplied with additional oxygen or not. Treatments with an oxygen mask alone, on the other hand, did not result in any significant improvements.

Two well-known and oft-cited studies by researchers at Harvard that found no connection between obesity and sleep apnoea, have since been withdrawn after one of the scientists, Robert B. Fogel, admitted to falsifying the data (Dolgin, 2009).

Although I never suffered from sleep apnoea myself, I did suffer from heavy snoring when I was overweight. This is now entirely a thing of the past, even when I sleep on my back. Once I reached a weight of about 120 kg, sleeping on my back was no longer an option, as it gave me backache, made me snore loudly, and later, caused the fat on my chest to press into my throat, closing off my windpipe. These days I need significantly less sleep in order to wake up refreshed in the morning, which has been a welcome, if unexpected, side effect of losing weight.

Arthritis/joint problems

The connection between excess weight and joint problems is pretty obvious since the joints are exposed to mechanical wear and tear, which is, of course, directly influenced by the weight they have to support. Depending on the movement involved, the pressure can reach up to four times a person's body weight, which is the case for the pressure that climbing stairs puts on the knees, for example. Every extra kilo means four more kilos of strain for the knees.

According to Guh et al. (2009), the probability of needing a replacement joint increases by 176 per cent for overweight men, and by 320 per cent for obese men. The risk increase for women is less extreme, at 80 per cent for overweight females and 96 per cent for those who are obese. The Harvard Medical School points to a study that found that more than half of overweight people suffer from acute knee pain, as compared to 15 per cent of the population of normal weight.

Gudbergsen showed in 2012 that losing weight was very effective in relieving the symptoms of obese patients over 60 with osteoarthritis of the knee. Significant weight loss was achieved by subjects in a 16-week diet program that began with very low calorie intake (around 500 to 800 kcal per day) and continued with 1200 kcal. The resulting weight loss led to pain reduction and improvements in day-to-day functioning of the subjects' knees.

As I said in the foreword to this book, it was problems with my knees that were the trigger for me to finally lose weight. The damage to my meniscus was repaired surgically and has not been a problem for me since. This means that I can't personally vouch for the extent to which losing weight helped get rid of my knee pain. What I have definitely noticed, though, is that my previously unstable knee (missing cruciate ligament) is now significantly steadier. Whereas I used to experience pain three to four times a month whenever I moved my knee in the wrong way, I have not had any problems with it now for several months. I think this is due partly to the fact that I have now built up the muscles around my knee, and partly to the fact that it is much easier to keep 65 kg upright and steady than 150 kg, which means I now have much better body control. As far as my increased risk of arthrosis (joint disease) is concerned, I can only hope that the combination of losing weight, building up muscle, doing physical exercise, and eating a high-protein diet will be enough to give me sufficient protection to prevent any damage and/ or defer the onset of arthrosis for as long as possible.

Sex/reproduction

Fatty tissue is not only an energy store, but also an organ that produces sex hormones. Both having too little and too much fat can therefore disturb the body's hormonal balance and cause serious problems. And those problems don't start in adulthood: one study of 2000 girls (Bralic et al., 2012) found that being overweight was associated with early menarche (an early first period). Crocker et al. (2014) examined boys and girls and found that being overweight was associated with higher oestrogen levels in both sexes. Breast development and pubic-hair growth was increased in overweight girls, while overweight boys were found to have smaller testicles and less pubic-hair growth.

Rich-Edwards et al. (2002) observed a U-shaped association between BMI and the risk of fertility problems. The lowest risk was found to be for women with a BMI between 20 and 24. Grodstein et al. (1994) found a strong increase in the risk of infertility among obese women and a slight increase in that risk for women with a BMI of below 17 and above 25. It appears that abdominal fat is particularly strongly associated with fertility problems among women (Norman et al., 2004).

Even when a pregnancy has begun, the risk of miscarriage and the death of the child increases if the mother is overweight or obese. A recent study (Aune et al., 2014) of almost 50,000 miscarriages, stillbirths, and infant deaths found that even a moderate increase in BMI increases the risk considerably. In 10,000 pregnancies, the absolute risk of foetal death for pregnant women with a BMI of 20, 25, and 30, were 76, 82, and 102; the risk of stillbirth was, 40, 48, and 59; the risk of the death of the baby within the first 28 days was 20, 21, and 24; and the risk of infant death was 33, 37, and 43, respectively.

Wilson et al., (2015) analysed umbilical cord blood and found that children of obese mothers had a weaker immune system than the offspring of mothers of normal weight. This might explain why the children of obese mothers are more likely to suffer from chronic inflammatory conditions such as asthma and cardiovascular disease.

Research by Stirrat & Reynolds (2014) showed that being overweight during pregnancy increases the probability of the offspring suffering from cardiovascular disease in adulthood by 15 per cent. If the mother is obese, that risk rises by 29 per cent.

According to a study by Tanda et al. (2012), the children of mothers who were already obese before their pregnancy perform significantly less well in literacy and numeracy tests later in life than the offspring of mothers of normal weight. Earlier research had already found that obesity in pregnant women has a negative effect on the foetal development of all the child's organs, because that development is very sensitive to environmental influences. One investigation of the development of the brains of the children of mothers who were of normal weight while pregnant and of those who were obese but otherwise healthy found that excess weight in the mother was associated with less-developed brain matter in their children (Ou et al., 2015b).

Overall, this means that for women who want to have children, being overweight reduces the chances both of getting pregnant and of having a healthy baby. The team of researchers Kort et al. (2014) showed that a 'meaningful' weight loss of at least 10 per cent of body mass increased the conception rate for overweight women from 54 to 88 per cent, and raised their chances of giving birth to a living child from 34 to 71 per cent. Clark et al. (1998) also found that an average weight loss of around 10 per cent achieved impressive improvements in infertile women: 60 of the 67 women saw a resumption of ovulation and 52 became pregnant. The miscarriage rate of 75 per cent sank to 18 per cent following the weight loss program. None of these positive effects were recorded in the control group of women who did not lose weight. This is why the first step doctors take in treating overweight infertile women is to advise them to lose weight.

Before the men start to feel left out, the Harvard Medical School published an article on obesity which makes reference to a 2007 study that found that each additional BMI point (around 3 kg in weight) was associated with a 2 per cent drop in testosterone levels. Also, abdominal girth represents a greater risk of reduced testosterone levels than age. Too little testosterone can lead to loss of libido and erectile dysfunction, as well as mood swings, exhaustion, weakness, problems sleeping, and lack of concentration.

Men with an abdominal circumference of more than 106 cm already have twice the risk of erectile dysfunction of men with an abdominal girth of 81 cm. A slightly overweight BMI of 28 increases the chances of suffering from erectile problems by 90 per cent.

But the situation isn't hopeless, as losing weight and doing exercise cured 30 per cent of men with erectile dysfunction, with no need for medication, while only 6 per cent of the men in the control group saw any improvement.

It seems that the results are less clear-cut when it comes to male fertility: while Chavarro et al. (2010) found that BMI has little influence on semen quality, except for a BMI of more than 35, Hammoud et al. found in 2008 that the risk of a lower sperm count rises from 5.3 to 15.6 per cent for obese men, and that the risk of reduced sperm motility rises from 4.5 to 13.3 per cent.

Another problem that can affect obese men, and which is less funny in real life than the many jokes about it might suggest, is the fact that as a man's weight increases, at some point his belly fat will eclipse his genitals, making his penis no longer visible. According to an article published in *The Evening Standard* (2012), this is already the case for a third of British men: they simply cannot see their own genitals because of their bulging bellies. The information website www.small-penis-facts.com points out that obesity can make a man's penis appear several centimetres shorter, as it is 'encased', so to speak, in lower belly fat. The good news is that losing weight can result in an additional 2 cm of penis length.

Asthma
It's equally unsurprising that there is a link between asthma and excess weight, seeing as an enlarged heart leaves less

space in the chest for the lungs to expand. The ribcage is also squashed from the outside by upper body fat, making breathing difficult.

Guh et al. (2009) found a 20 per cent increase in asthma among overweight men, and an increase of 43 per cent among obese males. Those figures for women were 25 and 78 per cent, respectively.

Here, again, losing weight is associated with positive effects. In a study of 14 severely asthmatic patients over eight weeks, a significant calorie reduction (to below 500 kcal per day) led to a fall in their average BMI from 37 to 32, along with great improvement in their asthma symptoms (Hakala et al., 2000).

A systematic analysis of weight-loss studies and asthma by Eneli et al. (2008) revealed that in every study examined, losing weight led to a significant decrease in the severity of asthma symptoms.

Back pain

Excess body weight puts strain on our bones and joints, which are not designed to carry more than the normal weight for long periods of time. They adapt to the strain as best they can — our bones get denser, for example — but the spine, with its delicate construction of vertebrae, intervertebral discs, and ligaments, is limited in its capacity to adapt its structure, making it more difficult for the spine to deal with excess weight than, for example, the thigh bones. Excess weight in various parts of the body, such as the chest or the belly, alters the body's centre of gravity, putting the spine of an overweight person under more strain than the spine of a person of normal weight with natural weight distribution.

According to Guh et al. (2009), the risk of chronic back pain increases by 59 per cent even for slightly overweight people, while the risk rises to 181 per cent for obese people.

The problem begins in childhood: in a comparison between obese children and children of normal weight, the former reported experiencing back pain more often and more intensely than the latter (Tsiros et al., 2014).

A study of gastric-band patients revealed that 58 per cent of obese people complained of back pain. After their massive weight loss following gastric-band surgery, only around a third of those patients still suffered from backache, and their pain was generally less severe, meaning it could be treated with lower levels of medication. Two-thirds of the previous back-pain sufferers were completely symptom-free after losing weight (Melissas et al., 2003).

In my case, I found that as my weight continued to increase, so did the amount of back pain I suffered. I still considered that I didn't really have a problem with backache, because I was laid up with severe pain once or twice a year, *only*. It wasn't until I'd lost the weight that I realised that I'd had a slight but permanent problem with back pain all along, and that it had restricted my mobility. Without realising it, I had adopted certain postural strategies to avoid pain when bending down, which I only noticed when they were no longer necessary. Sleeping in certain positions (on my back was impossible, on my front was only possible for short periods of time) meant that I was guaranteed to wake up with twinges of light-to-medium back pain in the morning, and I had unconsciously adapted my behaviour to this circumstance. Apparently, I have a natural inclination to sleep on my back, which I didn't realise for years. As you can see, my day-to-day life was definitely affected by back pain, but at the time it was the normal state of affairs, so I didn't even notice the symptoms I was suffering.

Gallbladder problems

The gallbladder helps in the digestion of fats. According to Guh et al. (2009), the risk of gallbladder disease is increased by 63 per cent in overweight males and by 151 per cent for obese men. For overweight women, the risk is 44 per cent higher, and it's 132 per cent higher for obese women.

The risk of developing gallstones is now increasingly affecting children and adolescents. ScienceDaily (2012) reported the findings of a study according to which overweight children have double the risk of developing gallstones. For obese and severely obese children, that risk was increased by six and eight times, respectively.

According to Erlinger (2000), the risk of gallstones for obese people in general is six times higher than normal, and the chances of developing new gallstones are increased by 10 to 12 per cent by highly calorie-reduced dieting. Very low-calorie diets usually include very little fat, but the gallbladder is sensitive to too little fat as well as too much. The risk of developing gallstones is up to 30 per cent higher in the first 12 to 18 months after gastric bypass surgery. In fact, the gallbladder is often removed during the same operation as a preventive measure.

I myself had regular attacks of biliary colic (bile is the fluid stored in the gallbladder), which is the severe pain when a gallstone is squeezed out through the narrow bile duct. When I began to lose weight, those attacks of colic disappeared completely. When I started eating more again, they returned. I ate a very great deal of fatty food over the 2014 Christmas holidays, and that led to a gallbladder inflammation, eventually requiring surgery. It's difficult to say in retrospect to what extent I could have avoided it, but it's generally the case that all attacks of biliary colic involve the risk of the gallbladder becoming inflamed. It could just as easily have happened to me during one of the attacks I'd suffered over the previous ten years, but then I would have been in much more danger on the operating table.

National Post (2015), reported that one in five severely obese Canadians who require emergency but routine surgery — such as removal of the appendix or gallbladder — don't survive the operation. (Severely) obese patients have the highest rate of complications and the longest recovery times, and a third have to undergo follow-up operations. Unfortunately, I couldn't locate the study that formed the basis for that article, but I did come across one by Ferrada et al. (2014) that found that there was only a slight, insignificant increase in mortality after surgery among overweight patients (BMI higher than 25) compared to patients of normal weight. However, that study did find that overweight patients ran twice the risk of post-operative wound infection, and they were more likely to need admission to intensive-care facilities after their operation.

Incontinence

One of the better-known conditions associated with excess weight and obesity is incontinence, which can be caused or exacerbated by excess weight exerting pressure on the abdomen (Cummings & Rodning, 2000).

In a study by Subak et al. (2002) of ten obese women who managed to reduce their BMI on average from 37 to 32, cases of incontinence were cut by half among those who lost more than 10 per cent of their body weight. A 2005 study (Subak et al.) of 48 women who had reduced their body weight by an average of 16 kg was similar. They managed to reduce their symptoms by as much as 60 per cent. For this reason, the authors of the study recommend weight loss as the first step in treating incontinence.

Although I was lucky enough never to have problems with incontinence, I can certainly imagine that being overweight has a big influence on the condition. I've noticed that I have to go to the toilet far less often now, where I used to have a real 'weak bladder'. I assume that my bladder just has more space now. My abdominal fat probably had a similar effect to the one pregnant women often talk about: your bladder gets squashed and can hold less fluid, meaning you have to visit the toilet more often.

Depression

The connection between depression and obesity is a complex one, and the relation between cause and effect is still the subject of a lot of research. In its obesity prevention recommendations, the Harvard Medical School points to a newly published, 28-year long-term study involving more than 58,000 people. According to that research, participants who were obese at the start of the study had a 55 per cent higher risk of developing clinical depression. At the same time, people who were already suffering from depression at the start of the study had a 58 per cent higher risk of becoming obese. The reasons for this link might be biological in nature, for example inflammatory or hormonal processes, but they might well also include socio-cultural factors.

According to research carried out by Onyike et al. (2003) the link between excess weight and depression is very marked among women, whereas it hardly exists among men. Extremely obese women have approximately four times the risk of becoming depressed than normal.

Interestingly, those results are complemented by the findings of Carpenter et al. (2000), who discovered that the relationship between weight and depression is turned on its head in the case of men. While the risk of becoming depressed was found to be equally high for obese men and those of normal weight, it was significantly higher among underweight men. For women, on the other hand, both being underweight and being overweight were associated with a higher risk of depressive disorders, although the highest risk was registered among overweight women.

Studying the effect on depression of losing weight is a difficult undertaking since changes in appetite and body weight are known to be symptoms of depression anyway. Reduced general motivation usually significantly restricts the activities of people with depression, meaning they generally burn less energy. At the same time, depending on individual disposition, the motivation to eat can be dulled by depression, causing extreme weight loss. Patients with a tendency towards emotional eating, by contrast, can gain a lot of weight because of depression. This means that research results are often skewed, as it's difficult to distinguish between intended and unwanted weight loss, precisely because extreme, unintentional changes in body weight are associated with depression.

Studies have shown that intentional weight loss has a positive effect on the symptoms of depression. One such study followed patients who underwent stomach reduction surgery: two years after their operation, the participants had lost huge amounts of weight. Before their surgery, patients scored an average of 10.6 in a depression test — the precise threshold score indicating the onset of depression. Two years later, their average test score was 4.4 — well within the healthy range (Strain et al., 2014).

An analysis of several studies since 1950 examining weight loss and depression revealed that intentional weight loss on average did reduce the symptoms of depression (Fabricatore et al., 2011).

At the time of writing, an article appeared online (FEELguide, 2015) reporting that new research had revealed that depression may be (partly) caused by inflammation. One researcher is quoted as saying, 'Inflammation is our immune system's natural response to injuries, infections, or foreign compounds. When triggered, the body pumps various cells and proteins to the site through the blood stream, including cytokines, a class of proteins that facilitate intercellular communication. It also happens that people suffering from depression are loaded with cytokines.'

The article goes on to say that adding anti-inflammatory medicines to antidepressants improves symptoms. The authors point out that this theory is also supported by the fact that healthy people can become temporarily depressed after receiving an inflammatory vaccine. As explained earlier, the normal and routine death of fat cells leads to a permanent state of inflammation in the bodies of overweight people, which might go at least some way towards explaining the possible biological mechanisms involved here.

Embolism/thrombosis

This is another area which has prompted a lot of research confirming the link between venous thromboses and being overweight or obese.

For example, Guh et al. (2009) found a 91 per cent higher risk of these conditions for overweight people, and a 251 per cent increase in the risk for obese people. Breaking those statistics down by gender, Stein et al. (2005) found obese men's risk of developing venous thrombosis was increased by 150 per cent, and their chances of suffering a pulmonary embolism were raised by 121 per cent. The risk of thrombosis was 175 per cent higher for obese women, and their risk of embolism was 102 per cent higher. Among both men and women under the age of 40, a high BMI was the most important risk factor.

Stroke

According to Guh et al. (2009), the risk of suffering a stroke is increased by 23 per cent for overweight men and by 51 per cent for obese men. The figures are 15 per cent and 49 per cent, respectively, for women.

The risk of a stroke as it corresponded to BMI and change in body weight was investigated in a 16-year long-term study involving 100,000 women. Compared to women with a BMI of less than 21, the risk for women with a BMI between 27 and 28 was 75 per cent higher. For women with a BMI of more than 32, that increase was as much as 137 per cent. An increase in body weight of between 11 and 20 kg after the age of eighteen was associated with a 69 per cent increase in risk of suffering a stroke. For women who put on more than 20 kg, the risk was increased by 152 per cent.

Gout

A higher body weight is strongly associated with the development of gout (Saag & Choi, 2006). Compared to people with a BMI of 21 to 22, the probability of developing gout increases by 40 per cent even in the upper-normal weight range (BMI 23 to 24), by 135 per cent for overweight people, and by 226 per cent for class I obesity. For class II obesity (BMI 35 plus), the risk is increased by 341 per cent.

According to Choi et al. (2005), losing 4.5 kg of body weight significantly reduces the risk of gout.

During my weight-loss process, my uric acid levels rose, which is associated with an increase in the probability of an attack of gout. I was told that this had to do with the rapid breakdown of my body's fat reserves. Unfortunately, I was unable to locate any studies to confirm this. All I managed to find was a mention in a study on the effects of very low-calorie diets (less than 500 kcal per day), but only eight of the more-than-4000 participants in that study complained of a worsening of their gout symptoms (Saris, 2012).

Nevertheless, this shows how important it is to undergo regular medical tests while losing weight. Luckily for me, taking medication to lower my uric acid levels while I was dieting meant that I didn't have a problem with gout. I've since been able to stop taking that medication.

Brain function

For me, and I'm sure for many readers, the most surprising and shocking discovery is that being overweight affects not just the body, but also the brain.

Evidence for this is provided by the results of a study by Gunstad et al. (2010), for example, which subjected 150 obese participants, two-thirds of whom were about to have gastric bypass surgery, to cognitive testing. Compared to the scores attained by healthy test subjects of normal weight, the obese participants' scores placed them at the lower end of the average range or even below average. Three months later, the subjects were tested again. Those who underwent gastric surgery had lost an average of around 22.5 kg. In the second round of testing, the surgery patients achieved average or above-average scores. The scores of the obese subjects who did not undergo surgery and had not lost weight were even lower at the end of the intervening three months.

A 2010 study (Kerwin et. al) tested the memory of 8745 women aged between 65 and 79. An increase of one BMI point was found to be associated with a drop of one point on the test's 100-point scale. The researchers hypothesise that inflammations due to fatty tissue may be responsible for damage to the memory.

Harvard's obesity-prevention report states that the risk of developing dementia is also closely associated with BMI. While underweight people run a 36 per cent higher risk of dementia compared to those of normal weight, the increase in risk for people who are overweight is 42 per cent.

In experiments with mice, researchers were able to prove that fatty tissue can impair the functioning of the brain. They found that overweight mice immediately began performing better in cognitive tests after their excess fat had been removed by liposuction. In contrast, lean mice who were injected with fatty tissue performed worse in the tests. The scientists found that the area of the mice's brains which is responsible for learning and memory was inflamed and swollen, and levels of certain neurotransmitter substances responsible for healthy communication between brain cells were reduced (NewsMic, 2014).

In their study involving more than 2000 people, Janowitz et al. (2015) found that a larger waist circumference was associated with a significant reduction in grey matter. The parts of the brain that were affected included the area responsible for the feeling of satiety. However, since this was only a correlational study, the authors were unable to make any conclusions about whether the reduced brain matter was caused by excess weight or excess weight was caused by the impairment of the sense of satiety.

This influence on brain function is active even in children. Chaddock-Heyman et al. (2014) say that children who are more physically fit have more brain matter and perform better in intelligence tests than their less fit peers. Ou et al. (2015a) also found evidence of a significant reduction in grey matter in obese but otherwise healthy children aged between eight and ten years. The health portal WebMD (2006) reported that children who were already very obese by the age of four achieved almost 30 points less than children of normal weight in IQ tests later in life, at the age of 11. The study pre-excluded children whose extreme obesity was due to Prader-Willi syndrome, which is associated with intellectual impairment. Those children had an average IQ of 63. The children whose obesity could not be explained by such a medical condition had an average score of 77, while their non-obese siblings scored an average of 106 in the IQ test. MRI scans also revealed differences in the structure of the children's brains. The scientist involved in the study pointed out that early infancy is a period in which the brain is still very much developing and is very vulnerable and therefore at its most susceptible to the damage caused by obesity.

But it's not only the brains of children that are more vulnerable: with increasing age, obesity again becomes a risk factor, this time for Alzheimer's. Dr Jeff Cummings, Director of Cleveland Clinic's Lou Ruvo Center for Brain Health, claims that obesity causes an increase in levels of certain proteins which lead to the development of the disease. Also, the obese have less brain mass, which is another risk factor. So it is, Cummings goes on, that although around half the risk of developing Alzheimer's is due to factors which we cannot influence, such as our genes and the ageing process, the other half is up to us: maintaining a normal weight and doing regular exercise reduces the risk significantly (Cleveland Clinic, 2013).

Ageing

Okay, first of all, ageing isn't a disease. But when cells age prematurely, it's certainly not a sign of good health. In a review article on telomeres, that is, the ends of chromosomes, Shammas (2012) shows how they grow shorter as we age. This shortening leads to cell ageing and cell death. So that means that shorter telomeres indicate an increased probability of illness or death. Shorter telomeres can be seen as a sign of wear and tear on cells. Telomere shortening is linked to many kinds of illnesses associated with (cell) ageing — such as diabetes, cancer, or heart failure, for example — and older people with shorter telomeres have a much shorter life expectancy than those with longer telomeres.

Ageing inevitably shortens telomeres, but lifestyle also has a great influence on their length. Smoking shortens them, for instance, and thus accelerates the ageing process by several years. Environmental pollution and stress also cause shorter telomeres.

Some of the biggest factors in the destruction of telomeres appear to be certain substances which are formed in fatty tissue. So, fat actively shortens telomeres, and in fact, this seems to be one of the most powerful factors for telomere length. Calorie reduction also appears to affect telomere length: experiments have shown that reducing food intake can drastically increase life expectancy in animals. Regular physical exercise also seems to guard against early or accelerated telomere shortening.

Additionally, a study involving mice (Xu et al., 2015) that were bred to store a lot of damaging fat tissue in their bodies and have a tendency towards early ageing and metabolic conditions such as diabetes, showed that calorie restriction was able to cancel out even those inbred negative effects and prevent premature ageing.

So, body weight and lifestyle really do have a direct influence on the way our cells age.

At least fat people are less likely to get osteoporosis

It's true: the risk of breaking a bone is almost twice as high with a BMI of 20 as with a BMI of 25, and the risk is even smaller for heavier people (De Laet et al., 2005). One of the reasons for this is that bones adapt to constant strain, becoming thicker and more robust. Nonetheless, this is not a zero-sum game: no one is forced to decide between the benefits of a lower risk of heart disease, diabetes, back and joint pain, gallbladder disease, gout, strokes, thrombosis, asthma, and incontinence, and the benefit of a lower risk of bone fractures. And your bones can also be strengthened with weight training (Nelson et al., 1994). It isn't necessary to carry a permanent extra load of 50 kg of body weight around — two-to-three strength training sessions a week are enough. Weights can be left behind in the gym at the end of the session, and your body isn't forced permanently to supply that extra mass.

According to the obesity paradox, overweight people have a better survival rate for some diseases

This theory makes me want to clutch my head in despair. Literally hundreds of previous studies showed that it's damaging to be overweight. Now, it turns out that, of people who are already sick, those who are overweight have a better prognosis than those with a normal or low BMI.

This phenomenon was first noticed in the treatment of cardiovascular disease, diabetes, and kidney failure. According to Oreopoulos et al. (2008), the mortality rate among patients with chronic heart failure is 12 per cent lower for overweight people, and 8 per cent lower for the obese, compared to people of normal weight. Romero-Corral et al. (2006) reached similar conclusions for patients with coronary heart disease. According to their research, slightly overweight patients had a 12 per cent lower chance of dying and the best chances of survival.

There are plenty of other studies on the 'obesity paradox', but to be honest, I don't understand what's supposed to be paradoxical about it. Is it really so surprising that someone whose body is damaged due to genetic or other factors should be worse off than someone whose otherwise healthy body has been badly treated for decades? That's like being surprised that a solid steel construction that has withstood decades of excessive strain is nevertheless more robust than a dam built by children out of sticks and twigs. To conclude from that that excessive strain must be a good thing is pretty far-fetched.

To illustrate just how absurd it is, as well as the obesity paradox, there's also something called the 'smoker's paradox', found in the same areas of medicine. It has been shown that smokers have better post-heart-attack survival rates than non-smokers. Even the fact that smokers are on average ten years younger when they suffer a heart attack does not fully explain their better survival rates. The smoker's paradox still exists, and there is no comprehensive explanation for it (*Live Science*, 2011). But the smoker's paradox serves to negate arguments in support of the obesity paradox such as 'maybe it's due to the extra stores of energy'.

Like obesity, smoking also slowly destroys a relatively healthy body, and so it is no wonder that such a body has better chances of survival than one that was born with defects. The only interesting thing about this is that the absurdity is much more obvious in the case of smokers than in case of body weight.

Your set point determines your weight

The so-called set point was still part of the scientific canon when I learned about it at the nutritional-science-based high school I attended. The idea is that at some point, energy intake and the body's biological system will find a natural equilibrium point and settle.

At the start of this book, I explained that the body's energy requirements increase as its mass grows. Let me give you an example: When I first met my husband eight years ago, I weighed about 80 kg. As I am 175 cm tall, my energy requirement was around 1900 kcal per day without much physical activity. Unfortunately, I was eating approximately 2100 kcal a day, leading to a surplus of 200 kcal, so I was gaining about a kilo a month. The heavier I got, the more energy my body required to function. At a weight of 90 kg, my body already required 2000 kcal per day, and eventually, just below that lovely 100 kg mark, my body's daily energy requirement was around 2100 kcal.

According to the set-point theory, my constant weight gain should have stopped at that point, since the number of calories my body was burning was about equivalent to the amount I was eating. For some people, that's actually what happens. Their weight settles somewhere around 'a bit chubby' without them feeling that they have to particularly restrict themselves in what they eat. To attain their ideal weight, they'd first have to eat fewer calories and then, to maintain that weight, they would have to eat less than before over the long term.

Unfortunately, that's not what happened in my case, and when I look back at the development of my weight, I see a continuous increase of 1 kg per month over a period of about seven years.

In the environment we live in, where tasty, high-calorie food is constantly available, it's not surprising that portions are increasing all the time. A recent study (Schoeller, 2014) found that people tend to put on an average of about half a pound over the winter holiday season, and most don't manage to shed it again. This leads to continuous weight gain over the years.

The set-point theory is interesting, but these days it's applied completely erroneously, and even when it's applied correctly, it isn't really relevant in practice because circumstances change all the time. For example, if I move from a first-floor flat to one on the third floor, or if I get a new job and have to drive to work rather than cycling, that will have an effect on my caloric balance. Even small changes to a daily routine can have a major influence. If a new coffee shop opens up on my route to work and I start buying a latte to go every morning, that can lead to a couple of extra kilos over a period of months. So, the set point is changing all the time and can only ever be a kind of estimate of what our body weight would be if we stuck to exactly the same routines for the next months or years.

The way many people use the term 'set point' is completely distorted. They use it to refer to some kind of genetically or otherwise predetermined weight which the body 'strives' for, because it wants to reach its 'natural weight'. Some people really believe that their body 'wants' to achieve a state of morbid obesity and weigh 150 kg.

The bottom line is that the set point is not the body's natural weight, but rather the theoretical final weight we would end up with if we continued in our exact same current eating and physical-activity behaviour patterns.

The 60 per cent of the population who are overweight can't all be 'lazy' and 'weak-willed'

I, too, don't believe that the reason people are overweight is that they are lazy or weak-willed. I got fat at the same time as completing my doctorate in psychology at the age of 25, and then I completed the very difficult psychotherapist training procedure while working full-time. I may be a terrible procrastinator with a tendency to dither, but I would never have achieved everything I have without putting in a certain amount of effort.

I don't think the statement above is fat logic exactly, but it's usually followed by a claim that the reason so many people are overweight is not (too much) food, but all manner of other things — from pollution to industrial food additives, to hormones, or vaccinations. These theories are based on an assumption that people still eat the way they did in the past, when the proportion of the population who were overweight was 10 per cent or lower, and only one person in every hundred reached the level of obesity. But now, goes the reasoning, despite similar eating habits and physical activity levels, some mysterious cause is making the majority of people fat.

The documentary film *Forks over Knives* by Lee Fulkerson sums it up very well: we come from a long line of ancestors who had to survive times of famine and needed to build up energy reserves whenever the opportunity presented itself. That's why we react so positively to the taste of sweet and fatty foods — and still today, our bodies try to prepare for future famines by keeping their energy stores as full as possible.

In prehistoric times, opportunities to build up energy reserves to the extent we do today were probably rare. And, although it isn't healthy to be carrying around 10 kg of extra fat, the drawbacks, which are chronic and only become noticeable after years, were less important than the advantages that those 10 kg of fat reserves brought in times of food scarcity. Ten kilos of body fat can allow a person to survive without food for months. And that was probably not an unlikely scenario for our ancestors, especially in winter. It would have made sense purely for reasons of survival for our forefathers to wolf down any bit of sustenance as soon as it was available. Thanks to modern inventions like deep freezers and supermarkets, we no longer need to carry our food reserves around with us all the time on our bodies. But evolution is a slow process, and our instincts have not yet adapted to this new set of circumstances.

Our instincts tell us we should cram ourselves full of as much fatty and sweet stuff as we can. Personally, I know very few people who truly have no natural appetite. Most of my thin friends and acquaintances like to overeat from time to time, just like anyone else. But the difference is that once they notice that the number on their bathroom scales is rising, or that their trousers are beginning to get a bit tight around the waist, they tell themselves they should cut down a bit — and then they follow their own advice.

Theoretically, or biologically, we would all have a set point of around 200 kilos if we really ate everything we possibly could.

We are one of the first generations to have to deal with this constant oversupply of food. Previous generations were much more physically active, and food was much less widely available to them.

Church et al. (2011), for example, estimate that daily work-related energy expenditure has decreased by more than 100 kcal over the past 50 years. More office jobs, increased automation, and less physical work mean that we expend less and less energy in our professional lives. What's more, *The Spectator* (2013) reported that three-quarters of the population used to walk for at least 30 minutes a day in the 1960s, while that applies to only 40 per cent today. This might also be due to the fact that nowadays more than twice as many households own at least one car as they did 50 years ago.

So, while our daily lives are becoming less and less active, we are confronted by an ever-increasing abundance and diversity of food. Even just the number of different types of chocolate bar on our supermarket shelves probably far outstrips the overall number of sweets available in the shops in the 1950s. On the same theme, many studies have shown that people tend to eat more when they are presented with a greater choice of food. For example, test subjects ate one-third more when they were offered sandwiches with a variety of fillings than when all the sandwiches were the same (Rolls et al., 1981).

Larger portions also encourage people to eat more. When test subjects were offered snacks, they ate more when 60 large pretzels were available than when they were provided with 120 pretzels of half the size. Similar results were recorded when different foodstuffs were offered (Geier et al., 2006). This fits with the results of an experiment by Nielsen & Popkin (2003), who compared portion sizes in 1977 and 1998, and found that portions had increased significantly, both at home and in restaurants, over those 21 years. A portion of savoury snacks had an average of 93 kcal more, for hamburgers it was an additional 97 kcal, for fries an extra 68 kcal, and beverages contained 49 kcal more per portion. Thus, as the portions we encounter gradually grow, we continue to eat more and more without even noticing it.

A large-scale meta-analysis (Hollands et al., 2015) of previous studies of portion size highlighted the significant effect that larger portions have on our calorie intake. The authors calculated that people ate between 144 and 228 kcal less per day when they were not confronted with overly large portions.

I think the reduced need for physical activity in our daily lives and the constantly available, rich supply of food are explanation enough for the gradual increase in body weight in our society. Guyenet (2014) calculates that our daily caloric intake has increased by 363 kcals over the last 50 years.

Unfortunately, there's still very little awareness of this in education. Even today, many children are pushed to (over) eat by parents who insist that they empty their plate ('…

because there are starving children in the world') or who constantly offer them snacks — a piece of chocolate here, a glass of juice there, even when they are not hungry or thirsty. In a society with an overabundance of food on offer, nutritional education should be focused on listening to the body's signals for hunger and satiety, and distinguishing them from pure appetite. Instead of being praised for finishing a portion, children should be encouraged to listen to their body when it tells them that they are full. Even if that means not emptying their plates at mealtimes.

I hope and believe that nutritional education will adapt to our altered circumstances in the coming decades and society will adopt a different attitude. It would be too much to expect such a profound change to take root in just a few short years.

Against this backdrop, I don't find it at all strange that more than half the population are already overweight and a large percentage are even obese. But the fact that it's not surprising doesn't change the serious consequences it will have, nor does it change the fact that we will have to 'relearn' many things — especially if we're going to get a good energy balance established in our habits.

We're all being taken for a ride by the diet industry

This statement is also not necessarily fat logic since much of the diet industry is definitely dubious or sells quack remedies. But on the other hand, and this is where fat logic kicks in, we are deceived into believing that the diet industry is a centralised social power that touts a completely absurd ideal that can only be achieved by special means like diet products, which earn the industry billions.

But first it should be pointed out that about half of all advertising is for food, drink, and stimulants. It is quite obvious that none of those branches of industry profits from people losing weight. Weight loss basically always brings with it a reduction in consumption. Industry wants us to consume its goods, and on those terms, the food and diet industries go hand in hand.

Companies that are trying to flog us snake-oil remedies or promote the latest summer-cabbage-soup-pineapple-low-carb diet are, of course, profit-driven and not particularly interested in our health. But the diet industry does not profit from sensible dieting, i.e., calorie reduction.

In principle, the food and diet industries have the same aim: to persuade us to consume as much food as possible, and to make us believe that food makes you happy and healthy, and that (too much) food is certainly not the reason we get fat. Instead, we can rely on remedies and secret tricks that allow us to carry on gorging ourselves and still lose weight. Don't consume *less*, whatever you do; better to consume more, *as well as* a product to help you lose weight.

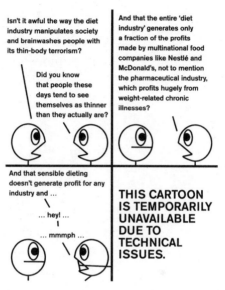

The last thing the food and diet industries want is for people to actually lose weight, and then maintain their figure, by eating less food every day. That would mean lower sales for them. The message that you can (literally) 'have your cake and eat it', that you can 'eat as much as you want' and still have 'the perfect body', is an attractive one. No one likes denying themselves things. That's why promises of 'better' and 'easier' solutions always find willing believers, even if they come at a cost. And if that dieting product fails to work, *Women's Weekly* or *Health and Beauty* will bring you news of the next pill or powder guaranteed to burn fat while you sleep or prevent your body from digesting more than 2 per cent of the calories you eat, and so give you your dream figure in an instant, with no effort on your part.

In this context, fighting fat logic means recognising the quack products of the diet industry for what they are and instead relying on an individual, sensible, and medically supervised weight-loss process.

I know someone who lost loads of weight on this new diet — it alters your metabolism, so you can eat as much as you want

Nutrition gurus regularly appear on the scene, promising that a certain kind of diet, such as Atkins, raw food, low-carb, or veganism, will affect the metabolism in such a way that you can eat as much as you want (within the constraints of their nutritional regime) and still lose weight with a daily intake of 5000 kcal.

Bullshit.

Bullshit with a grain of truth, to be exact. Certain regimes do actually lead to your *wanting* to eat less, which means that the claim that 'you can eat as much as you want' and still lose weight is sort of true. But it isn't magic, and it doesn't work for everyone, as some people don't let themselves be fooled and still consume more calories than they need, only now it's not in the form of cake, but chicken or broccoli.

Research shows that a high-protein diet curbs your appetite and that people consume an average of 440 kcal less when their food is high in protein (Weigle et al. 2005). One reason for this is that protein leaves us feeling full for longer and has less of an impact on blood sugar levels. While simple carbohydrates like sugar and white flour enter the blood stream rapidly, provide a lot of energy very quickly, and are followed by a 'crash', dietary fibre (e.g., from fruit, vegetables, or wholemeal products) cause sugars to be absorbed into the bloodstream at a slower and more constant rate, so they make us feel full for longer, thereby avoiding snack attacks and sugar rushes. Protein and fat are also absorbed more slowly and provide the body with energy for longer.

So while a piece of cake with 500 kcal and a piece of turkey meat with broccoli also containing 500 kcal will provide the same amount of energy, the cake is processed more quickly and delivers its energy straight away. This can result in pleasurable feelings, just like the consumption of coffee and nicotine, and even cocaine — all of which have a stimulatory effect. After the rush comes the crash, as we feel the sugar levels in the blood fall again. This is an unpleasant sensation, stoking the desire to do something to counter it — preferably with something that will rapidly raise our blood sugar levels again. The paradox here is that a diet containing lots of sugar and white flour products often leaves us feeling less contented, even though eating those delicacies is considered 'a treat'. They leave us with the feeling that we can never quite eat *enough* as we constantly strive to regulate our blood sugar at a higher level.

MedicineNet (2015) additionally explains that people whose blood sugar level is high for a prolonged period of time can eventually become so accustomed to it that they feel symptoms of hypoglycaemia when their sugar level sinks to normal. Their heart rate increases, they begin to tremble and feel restless, anxious, and dizzy. These are similar to withdrawal symptoms.

High-fibre foods such as fruit also contain a lot of sugar, and similarly, wholemeal bread often contains the same number of carbohydrates as white bread, but the fact that the energy is taken up more slowly means they don't cause sugar rushes or crashes. The energy is released in a more constant way.

High-fibre foods also have much more volume. A pound of carrots contains fewer than 200 kcal, and it's far easier to polish off one 50-gram Snickers bar (242 kcal) as a snack than half a kilo of carrots. If we assume that vegetables contain an average of 30 kcal and fruit has about 60 kcal per 100 g, then we see that it is physically almost impossible to overeat on raw foods. A large, muscular man with a daily energy requirement of 2500 kcal would have to eat 5 kg of fruit and veg per day to gain weight. Even for people with naturally large appetites, that would be quite a challenge.

The secret of super-diets is that they lead people to cut their calorie intake without feeling that they are denying themselves anything, because their calories now come from high-protein or high-fibre sources, which reduces their appetite. But that doesn't mean that the principle of caloric balance no longer applies. It is, of course, just as possible to lose weight by eating 1000 kcal of cake per day as by eating 1000 kcal of vegetables or protein.

This is proven by the case of two men, a teacher and a university professor, who both lost a lot of weight by eating 'unhealthy food' but always only up to a certain calorie limit. The teacher, John Cisna, ate 2000 kcal of McDonald's every day and went from 127 kg down to 101 kg in six months and lowered his BMI from a previously severely obese level to 31 (Pawlowski, 2014).

Mark Haub, a professor at Kansas State University, consumed 1800 kcal a day in the form of sweet snacks like doughnuts and Twinkies, and lost 12 kg in two months, taking him from being slightly overweight to within the normal weight range. Surprisingly, his blood test results improved, despite his unhealthy diet: his levels of 'bad' LDL cholesterol fell by 20 per cent and his 'good' HDL cholesterol rose by the same amount, while his blood lipid levels sank by 39 per cent (Park, 2010).

Larger-scale studies comparing different eating regimes showed similar results. Golay et al. (2000), for example, gave obese hospital patients a diet of 1100 kcal per day, either in the form of a low-fat diet or as a diet low in carbohydrates but high in fat. The two groups lost the same amount of weight, and both saw similar and significant improvements in their blood test results and their blood pressure, irrespective of the diet they had been fed. The same group of researchers had carried out a similar experiment four years earlier (Golay et al., 1996), and in that case, too, patients who were given a diet of 1200 kcal a day either in the form of a high-carbohydrate or a low-carbohydrate diet lost the same amount of weight. They also saw great improvements in their blood test results. The group on the low-carb diet showed a significant drop in their blood sugar levels, however, unlike the high-carb group.

The amount of sugar in a diet is also irrelevant, as long as the overall number of calories consumed remains the same. Surwit et al. (1997) placed 42 women on a high-sugar or a low-sugar diet to lose weight. Both groups lost the same amount, both saw improvements in their blood test results and their blood pressure, their appetites shrank over time, and both groups' general mood improved.

Healthy eating is just too expensive. It doesn't matter how much you eat, what's important is *what* you eat

Looking at the evidence from the previous chapter, it becomes obvious that the argument that healthy eating is expensive is not true. In order to lose weight, the crucial thing isn't *what* you eat, but *how much*. In theory, there's no need to change your diet as long as you stop eating as soon as your consumption reaches a certain number of calories.

Of course, what you eat affects how full you feel. Nickols-Richardson et al. (2005), for example, compared the effects of a high-protein, low-carb diet with those of a low-fat diet on older, overweight women. While both the low-fat and the low-carb group lost weight, the low-carb women reported feeling hungry less than their peers whose diet was low in fat.

A study by Johnston et al. (2014) that compared a number of popular diets came to a similar conclusion: while protein-based, low-carb diets had slightly better results as far as weight loss was concerned, all the diets worked in principle, as long as the dieters stuck to them.

When it comes down to it, it doesn't matter which diet you choose, whether it be Atkins, Weight Watchers, or calorie counting. As long as you stick to it and take in less energy than you consume, you will lose weight. Whether your 1200, 1500, or 1800 kcal per day come in the form of chocolate, chips, and burgers, or grilled chicken breast and broccoli is of secondary importance for the actual weight-loss process.

What's more, many low-calorie, high-protein foodstuffs, like low-fat cottage cheese, natural yoghurt, eggs, brown rice, and frozen or seasonal fruit and vegetables, are very cheap and are quick and easy to prepare.

95 per cent of all diets end in failure

The argument that '95 per cent of all diets end in failure' is often rolled out by the fat acceptance movement to suggest that for 95 per cent of people it's impossible to lose weight, and that those who succeed are statistical outliers who certainly can't be taken as proof that body weight can be controlled or changed.

The claim that it's fundamentally impossible for people to lose weight for physical reasons is 'proven' by statistics which say that 95 per cent of test subjects return to their original weight or exceed it within five years of dieting. It is difficult to trace the origin of this myth, as following the 'sources' quoted just bounces you from one blog to another and on to the next. Most of the studies that are quoted are very old, some dating back to 1959, for example, and involved a small cohort of test subjects in a clinical environment (*The New York Times*, 1999). These studies were carried out with people who were already considered to be 'difficult cases', and who had already tried everything to lose weight before being accepted into a clinical trial. Using studies like these as proof is like measuring the success of a maths program on the performance not of average students, but of those who attended courses for people with learning difficulties or who have dyscalculia. It might show that even intensive study does not usually lead to an A-grade performance in maths tests, and, in turn, that result might lead to the assumption that studying mathematics isn't an effective way to improve performance in maths.

To add to this, those early weight-loss 'programs' consisted of thrusting a diet plan into the test subjects' hands and sending them away to get on with it. It's hardly surprising under such conditions that only 5 per cent of participants managed to lose weight in the long term. It's like offering heavy chain smokers a program that consisted of just saying to them, 'Don't smoke any more cigarettes. Good luck! We'll see you in a year to find out how it went.'

In fact, more recent studies indicate much better success rates. An analysis by Anderson et al. (2001) of several studies showed that participants in weight-loss programs had maintained an average weight loss of 3 kg five years after taking part. That might only be equivalent to one BMI point, but it contradicts the argument that people are inevitably doomed to end up weighing more after a diet than before. Even if some participants do regain some weight, the general trend is encouraging.

Wing & Phelan (2005) claim that around 20 per cent of diet-program participants maintain a weight loss of at least 10 per cent of their original body weight.

Wadden et al. (2011) examined the weight loss achieved by type 2 diabetics over four years and found that 45 per cent managed to achieve and maintain a weight loss of at least 5 per cent of their original body weight.

For ten years, a long-term study by Thomas et al. (2014) observed people who had lost more than 14 kg. It found that 87 per cent still maintained a weight loss of more than 10 per cent of their original body weight after those ten years. This study was also one that finally examined 'normal' dieters, most of whom had lost weight without any medical assistance.

Another interesting study was authored by McGuire and team (1999), who carried out telephone interviews with people picked at random. Of those questioned, more than half (54 per cent) said they had lost more than 10 per cent of their body weight at some time in their lives, and the weight loss was intentional for about two-thirds of those people. Of those who had lost weight intentionally, around half said they had maintained their lower weight for more than a year at the time of questioning, and around a quarter had kept the weight off for more than five years. The good thing about this study is that the responders were picked completely at random, providing an accurate picture of the population at large: it concentrated neither on 'difficult cases' taking part in clinical trials, nor on people who had actively signed up as volunteers. In my view, this is the study that most accurately reflects the real success rate in losing weight.

It's difficult to find truly reliable data on the success rates of dieters. I know people who go on some diet or other every few months and keep it up for a couple of weeks — fasting or detoxing — to counterbalance the weight they gain in the intervening months. They consume a couple of hundred calories too many per day and, after gaining 3 to 4 kg, go on a two-week fast. Can those people's diets be said to be successful, or failed? It's difficult to pin down because, on the one hand, they have lost weight by dieting, but on the other hand, they always regain that weight later. In terms of their aim of maintaining their weight at a level they find acceptable, they are successful, even if they could also be said to have a failure rate of 100 per cent if their diets are viewed individually.

Of course, it's all a question of attitude. If you want to lose weight, you have to eat fewer calories than 'normal' for a limited period of time. What most people forget, though, is that they didn't get to be overweight by eating normally, but by eating more than normal. Someone who eats normally in the sense of 'according to what their body needs' after a diet, won't put the weight back on. But someone who eats normally in the sense of 'the same as before their diet' will naturally revert to their original starting point — but that can't be said to be the fault of the diet.

To say that 95 per cent of all dieters put the weight back on and to claim that this fact proves that diets don't work, is like saying the low success rate among smokers trying to quit (six out of every seven people restart the habit) proves that not-smoking doesn't work. Of course not-smoking works. If you never light up another cigarette again, you have successfully become a non-smoker. Our bodies don't produce yo-yo cigarettes that materialise miraculously between our lips of their own accord, just as they don't miraculously return to their original weight of their own accord. *People* return to their original weight.

The bottom line is that dieting works. Diets work. And 100 per cent of people are able to lose weight.

—

Yo-yo dieting is way more unhealthy than being fat

I have read this claim many times. But that doesn't make it true. In 1992, Jeffrey et al. investigated this very question. In their study involving 202 obese test subjects, they wanted to find out whether large fluctuations in body weight of more than 5.4 kg increase the risk of cardiovascular disease. Their results did not support that hypothesis, as the blood values of the continuously obese subjects did not differ significantly from those with large fluctuations in body weight. In seven subjects, the results went in the opposite direction to that hypothesised: the subjects with large fluctuations in body weight were healthier than the continuously obese subjects.

A review article authored by Mehta et al. in 2014 also points out that evidence for the adverse effects of such 'weight cycling' is sparse, if it exists at all. *NBC News* (2011) reported a study using mice which found that yo-yo diets are healthier than staying obese. The overweight mice had worse blood-test results and had only three-quarters the lifespan of the mice on the yo-yo diets. During their overweight phases, the yo-yo mice's blood tests were just as bad as the constantly overweight ones', but better during the periods when they lost weight. They lived almost as long as the constantly lean mice.

A very large long-term study by Stevens et al. (2012) also investigated whether yo-yo dieting impacts negatively on life expectancy. It came to the conclusion that studies that found an increased death rate in people with fluctuating weight did not differentiate between intentional and unintentional weight loss. Many life-threatening conditions, such as cancer, strokes, or diabetes, for example, often cause massive weight loss, leading studies which did not take this into account to find an increased risk of death with weight cycling. In fact, it is the disease that causes both the weight fluctuations and the eventual death. This isn't comparable in any way to healthy people who deliberately decide to reduce their body weight. In their study, Stevens et. al took account of these factors and discovered that having up to four episodes of weight loss in the past was associated with a lower mortality rate, but more episodes than that had no effect. Whatever the case, weight cycling was not found to be harmful.

But what about the psychological aspect? If yo-yo dieting doesn't pose a risk to physical health, and doesn't 'ruin the metabolism', might it still perhaps make yo-yo dieters less likely to lose weight and maintain it? No. Previous (yo-yo) diets in the past do not have a negative influence on future chances of success. Having lost weight a few times does not lower the probability of losing weight permanently, either for physical or mental reasons (Mason et al., 2013).

The researcher Dr Thomas Rüther of the Tobacco Addiction Clinic at Munich University Hospital notes moreover that smokers also fail to quit six times on average before succeeding at the seventh attempt (*TZ*, 2012). But each failed attempt is a step closer to success.

I think something similar holds for dieting. Previous attempts offer an opportunity to analyse where the problems lie, what works and what doesn't, and which strategies might be promising for future success. A detailed examination of the past with a view to identifying certain sticking points can help us find better ways to lose weight and keep it off.

One reader of my blog wrote to me to say how her previous attempts at losing weight had been hampered by the myth of 'starvation-mode metabolism'. She had always been told she had to eat at least 1200 kcal a day to avoid going into starvation mode. As a small woman with a rather sedentary lifestyle, that left her with only a slight deficit, so that just one treat on the weekend or one birthday party 'ruined' her results for the entire week. She eventually found herself in the paradoxical situation where she was forcing herself to eat on normal workdays, when she didn't have much appetite, just to avoid the 'starvation mode', and then on special occasions, like a dinner party, she felt she couldn't afford to let herself indulge. Abandoning the starvation-mode myth helped her take a more relaxed attitude to her own eating behaviour and she was able to eat as little as she wanted on days when she had no appetite, leaving her with the capacity to treat herself on the few occasions when she went out to eat, and still manage to lose weight as the weeks progressed. Hers is an ideal case, of course: she was able to identify a particular problem, and then alter her strategy to achieve success with relatively little effort. But ideal or not, it shows how 'failed' attempts to lose weight can actually be a key part of the process towards eventually losing weight successfully, and in this way, so-called yo-yo dieting is actually a step in the right direction.

You have to give your body what it wants. Constant restriction is unhealthy

I'm the last person to oppose a little indulgence, but I have to point out that this statement includes a pretty big portion of fat logic.

Proponents of 'intuitive eating' often claim that our bodies 'know' intuitively what they need — for example, if you feel the need to eat chocolate, that's because it contains some nutrient that your body currently requires. There may very well be some people out there who intuitively eat a healthy diet, but that doesn't take away from the fact that our bodies have evolved with a preference for sweet and fatty foods — that is, those with highest possible calorie content.

In the ancient past, when food was generally less abundant and there were no industrially processed products, that preference for sweet and fatty foods was a useful strategy. The foodstuffs with the highest calorie count were nuts, oily fish, or fatty meat, but they had to be laboriously gathered, caught, or killed. From an evolutionary point of view, it made good sense to develop a strong (taste) incentive to eat these foods as a source of important nutrients, rather than spending the entire day munching on masses of vegetables and berries.

Because of this, if our bodies got their way, we would almost always prefer a greasy burger or a bar of chocolate over a carrot or an apple. So, for someone who already has unbalanced eating habits and is therefore under- or overweight (or someone of normal weight with an unhealthy diet), 'intuitive eating' is comparable to 'intuitive drinking' in an alcoholic. The body's signals have long since become confused, and alcoholics do experience the physical feeling that they *need* alcohol, becoming restless with cravings, and even experiencing withdrawal symptoms like shaking, when they don't get a drink. Their bodies are so strongly habituated to the substance that their system has adapted to it. In a similar way, our system also adapts to an unhealthy diet over time, for example with changes in our gut flora.

When this happens, the body has to be consciously retrained so as to eventually regain a healthy intuition after some months or years.

I think the idea that intuitive eating is healthier is based on an inverted cause-and-effect relation: on average, people who have never had problems with their weight or their health are substantially more likely to eat intuitively than those who have, or have had, difficulties. Most people who have reached an unhealthy state (in terms of body weight or other problems) through intuitive eating, will at some point try to remedy the situation by taking countermeasures (dieting, nutrition plans, calorie counting), which means they can no longer be considered intuitive eaters.

To say that eating a calorie-controlled diet is bad is like saying that frequent visits to the doctor are bad because people who go to the doctors a lot are ill more often. But you don't get sick from sitting in the doctor's waiting room;

you sit in the waiting room because you're sick. It would be daft to tell someone with a serious illness not to see their doctor because there is a connection between visiting the doctor and sickness.

Concerning the claim that 'restriction is always bad': it's now been known for more than 70 years that long-term caloric restriction increases longevity in various animal species like rats, mice, hamsters, dogs, and fish. Caloric restriction slows down the ageing process and guards against chronic disease in old age. This doesn't appear to be a result of reduction in body fat, but of the caloric restriction itself (Masoro, 2005).

Carrillo & Flouris (2011) assume caloric restriction lowers body temperature, which has a positive effect on longevity because, among other things, it influences hormone regulation and prevents inflammation. The mechanisms by which caloric restriction slows down the ageing process are not yet sufficiently understood. But it does seem clear that both a lower proportion of body fat and restricted caloric intake have a direct, positive effect on physical health and longevity.

I'd rather enjoy life!

My decision to be fat was mainly based on two assumptions: one, being fat is not all that unhealthy, and two, being thin (for me) would mean massive restrictions, sacrifices, and self-denial.

This is an example of a typical day's consumption for me — around 1200 kcal. It begins with a large breakfast of four waffles, two large main meals, and two snacks between meals. I can't pretend that this nutrition plan has left me hungry or with feelings of self-denial. The internet is full of recipes for low-calorie meals, and even traditional dishes can often be altered to reduce their calories without making them any less tasty.

My salad, for example, which contained 1500 kcal, is now only 200 to 500 kcal, depending on how I decide to have it. The changes I made weren't that huge. I replaced the fatty salmon with chicken or turkey, the ball of mozzarella with half a ball of a much less calorie-laden, low-fat mozzarella, and the lettuce, tomato, and bell pepper remain unchanged. Instead of using three tablespoons of oil in the dressing, I now make it with water and a little sweetener. Those changes have hardly altered the taste — I still enjoy my salad just as much, even though it now contributes 1000 kcal less, easily, to my daily intake.

It's definitely worth seeking out these versions of dishes as well as trying out new meals. For my part, I had a lot of fun trying out new recipes. I often find that I enjoy my new meals more than I used to enjoy my thoughtless eating. Since becoming conscious of the calories I 'spend' on it, I have come to appreciate food more, and I experience the enjoyment more consciously.

I have now reached my target weight, and I do a lot of exercise. And, seeing as my metabolism hasn't been 'ruined' by losing weight, I can now consume about 2500 to 3000 kcal a day. That's about the same as when I weighed 150 kg. One positive side-effect of my losing weight is the fact that exercise, which I used to find so torturous, is now enjoyable.

On balance, that means that I now have more enjoyment in my life, rather than less. I look forward to every hour I spend on my elliptical trainer watching my favourite TV series, to every run I go on, every bike ride with my husband, or every gym date with a friend for weight training. For the first time in my life, I sometimes find myself reviewing the day and realising that I haven't eaten enough and could actually eat a bag of crisps or a bar of chocolate. It took over a year to reach this point, and it was a year in which sacrifice was the rule. But it wasn't what I had feared: I didn't have to choose between a happy life full of enjoyment and a skinny life full of sacrifices.

The fact that so many people equate being thin with not enjoying life is, I think, a great problem. Just like the clichés about weak-willed and lazy fat people, it suggests that normal body weight is something that can only be achieved and maintained through permanent self-discipline and a joyless lifestyle.

I now think that the opposite is true. Even though I had convinced myself that I was okay with my weight, food was always connected with a feeling of guilt for me. Whenever I ate chocolate or ordered a pizza, I was always aware that I shouldn't really be doing it. And my enjoyment was coloured by that niggling guilty conscience in the back of my mind. Paradoxically, the opposite is now the case. Counting calories doesn't make me feel *more* guilty, but less. I can consciously choose to eat a box of chocolates if I feel like it and enjoy it without regret because I know that it fits into my daily plan.

The funny thing is, I now often experience *the other side* of what I thought for many years. If I'm going out for a meal, for example, I plan things so that I can eat as much as I want. And when people see me going back to the buffet to pile up my plate with seconds, I sometimes get comments like 'You're so lucky, you can eat whatever you want!' People who don't know me from before assume I'm just blessed with a fast metabolism, and those who did know me are usually amazed that I still 'eat normally'.

It's an interesting experience to be perceived as a 'naturally thin person' or as someone 'blessed with a fast metabolism', after all those overweight years. It's something I used to yearn for — now I want to tell them, 'No! That's not true. My metabolism isn't special, and it has nothing to do with luck!'

You have to follow these certain rules to lose weight

From 'You have to eat six small meals a day to lose weight' and 'You must eat breakfast to kick-start your metabolism', to 'You have to eat at least 1200 kcal a day' or 'Don't eat carbs in the evening' — there are thousands of (nonsensical) rules about losing weight and/or eating healthily.

Ultimately, the only thing that counts is daily caloric intake. It doesn't matter whether the calories are distributed over six meals, 20, or just one. The 'six meals a day' rule has been disproved umpteen times but still stubbornly persists as a widespread myth. Bellisle et al. (1997) were able to show that the way food intake is distributed throughout the day (one large meal or several small ones) doesn't influence the body's energy requirements. And Cameron et al. (2010) showed that there was no significant difference in weight loss between dieters who ate six meals spread through the day and those who only ate one.

Aside from the weight-loss aspect, which is unaffected by the distribution of calories across the day, there are indications that eating (too) often throughout the day can have an adverse effect on the regulation of your blood sugar levels. The body gets used to a constant supply of energy, so to speak, and forgets how to regulate blood sugar by itself. By contrast, so-called intermittent fasting can have an extremely positive influence on blood sugar regulation.

The phrase 'intermittent fasting' is used to refer to either whole-day fasting, or time-restricted feeding, when eating is limited to one part of the day (for example, the first four hours of the morning), and the rest of the time (20 hours in this case) is spent fasting.

Funnily enough, the reaction to this kind of fasting from the outside world is often 'It can't be healthy!', which just goes to show once again how many of these eating rules continue to circulate unquestioned. They're sometimes even spread by so-called experts, although they aren't supported by any scientific research.

Whether you have breakfast in the morning or take your first meal of the day at 8.00 pm is irrelevant for losing weight successfully. It doesn't matter whether you eat carbohydrates or protein in the evening, and I've already commented extensively on the idea that you should never eat less than a certain amount per day. Of course, some of these interventions do have a short-term effect on the number that appears on the scales. If I eat a large plate of pasta in the evening, then I will naturally weigh more the next morning, because that food is still in my system. If I start eating carbohydrates again after having cut them out of my diet for a time, my body will retain more water — and the same is true of salty food. But I've already commented enough on the topic of short-term weight changes. When and how you eat is completely irrelevant for long-term weight loss.

Some rules are based on scientific research. But it should be pointed out that what counts as a difference in a scientific study can sometimes ultimately be no more than a few calories. Whether you 'kick-start' your metabolism in the morning or not might make a difference of ten calories, depending on whether you have your breakfast right after getting up or four hours later, and provided you don't change your daily habits in any other way. But these 'never skip breakfast' rules make some people feel forced to eat even if they don't have the appetite. In the worst-case scenario, they will end up eating more overall, because they feel breakfast somehow doesn't count, as it was more of a duty than a pleasure for them.

BREAKFAST LOGIC

WHAT THEY SAY:

It's very important to eat breakfast in order to kick-start your metabolism in the morning. After eating breakfast, you burn more calories than if you haven't eaten anything.

THE FACTS BEHIND IT:

It takes about 10 per cent of the energy contained in food to digest it. So after you eat something, you'll use energy (and burn calories) digesting it.

EQUIVALENT LOGIC:

It's important to go shopping if you want to save money. If I buy a pair of trousers that are reduced from 100 euros to 90 euros, I will have saved 10 euros more than someone who didn't buy any new trousers.

The Washington Post (2015) devoted an extremely interesting article to the science behind this breakfasting tip, explaining its origin. The recommendation entered the US Dietary Guidelines in 2010 on the basis of purely observational studies after it was noticed that people who regularly eat breakfast have a lower risk of obesity on average. But such observational studies are not proof of cause and effect. It could easily be the case that other factors are the cause of both effects — the habit of eating breakfast *and* lower body weight. Also, taking the fact that people who eat breakfast are more likely to be slim as the command 'Eat breakfast, then you'll be thin!' makes about as much sense as concluding that people who call the emergency services get injured more often, so to reduce your risk of injury, you should never dial the emergency number. The *Washington Post* article also cites several studies that proved by experimentation — not simply by observation — that eating breakfast definitely does not lead to weight loss, and that skipping breakfast doesn't make you fat. The article highlights how unfounded some 'official recommendations' are. A non-expert public unaware of how the recommendations come about, naturally assume there must be something to them. After all, they're official guidelines, so there must be something behind them — probably some kind of 'metabolism stuff', some 'hormone thing', or other mysterious influences.

The best strategy is to ignore all the tips and keep it simple: take in less energy than you burn. In the end, it's more important to find the way of eating that works best for you and that makes it easy (easier) for you to maintain your personal caloric balance. Your nutrition plan should not try to follow any general rules, but should rather reflect your individual needs. Some medical conditions can even be exacerbated by following certain rules. What works for some may be precisely the wrong thing for others with different (e.g., medical) backgrounds. For some people, it might be helpful to distribute their caloric intake across many small meals throughout the day. But if that's effective for their weight loss, it won't be for physical reasons, but mental ones. In a case like that, it makes sense for that person to eat many small meals, rather than forcing herself to follow the current trend of intermittent fasting.

I think a lot of people find it unsettling to throw all their 'golden rules' overboard. Some almost see it as an attack, because there will always be people who have had a kind of personal breakthrough by following certain rules. It might be avoiding carbohydrates (in the evening), following a particular diet, changing the frequency of their meals, or losing weight with some kind of shake or supplement. It is important to remember here that 'there's more than one way to skin a cat' and different people work in different ways. An easy route for some people might be tortuous for others, and vice versa.

For example, I, personally, like the structured and clear-cut nature of counting calories. I have a very strong tendency to lie to myself. ('I haven't really eaten that much today, so a piece of chocolate should be okay … Oh, I've already eaten too much chocolate today, so I may as well write the day off as a failure and eat some crisps!') Seeing the numbers written down in black and white and knowing for sure whether I can afford to eat that bar of chocolate helps me to avoid fooling myself. At the same time, I know there are people who can think of nothing worse than having to weigh out all their food and type the figures into an app, and who do a lot better when they tell themselves, 'I'll eat only lean meat, fruit, and vegetables six days a week, and one day a week I'll eat whatever I want,' or still others who would rather fast for two days a week, or do without dinner every evening, so that they *automatically* eat less, without the bother of counting calories.

Those who have achieved a breakthrough with a certain diet often tend to mystify their success as being down to some kind of magical, metabolism-accelerating effect or equivalent. That's usually because it really does seem like a miracle when you suddenly find a way that feels easy to solve a problem that you've been struggling with for years. In reality, there's nothing magical about those miracle diets. Each one is just one more way of reducing calorie intake, among myriad others. Even if it feels like you've found a trick that lets you eat as much as you like, the real trick is that you want to eat less, and so cutting back no longer feels like a sacrifice.

I don't think it can do any harm to look into and try out some of the diet fads, like a low-carb diet, the 5:2 diet, the paleo diet, or protein shakes as meal replacements, to see if any of them suit you personally. Ideally, there will be one among them that makes it relatively easy for you to lose weight and keep it off. For me, for example, a protein-based diet is good, both because I do a lot of weight training and therefore have a high requirement for protein, and because protein is more filling and therefore reduces my appetite. But I love sweets too much to follow a very strict protein-based plan.

So, while it might be disappointing for some people to see their golden rules debunked, I think it's actually extremely liberating. I know a lot of people who have a permanently guilty conscience about their eating habits — be it because they just can't manage to force down breakfast in the morning, often only have one big meal a day, eat too late in the evening, or like to eat the wrong carbohydrates at the wrong time. In these cases, it can be very freeing to cast off that guilt and sit down at 8.00 pm with a plate of pasta in the knowledge that it absolutely fits in with your daily calorie requirements, and it won't automatically go straight to your hips just because you're eating it five hours 'late'.

If losing weight was really that easy, everyone would be thin

'If not-smoking was really that easy, and all it took was just to decide *I won't smoke another cigarette (ever again)*, there wouldn't be any smokers.'

'If not-drinking was really that easy, and all it took was just to decide *I won't drink alcohol (any more)*, there wouldn't be any alcoholics.'

Just because something is easy said, it doesn't mean it's easy done. The physical aspect of losing weight is actually really simple, but it's often turned into something incredibly complicated, and a lot of people seem to spend their whole lives on an epic quest to find the holy grail of the miracle diet that will help them lose weight permanently.

Some pearls of fat-logic wisdom are comparable to telling an alcoholic he can't become a teetotaller just by stopping drinking, because not-drinking will increase the amount of alcohol in his blood. So, he should drink at least three beers in the morning, but definitely not after 1.00 pm, and it has to be dark beer, unless his blood group is O or AB, in which case it has to be lager. Also, he must have a blood alcohol level of at least 2 parts per mil, or his alcohol metabolism will be ruined, and he'll become permanently drunk.

All this pseudo-scientific nonsense is seductive, though, because the more complex and complicated achieving our desired aim is seen to be, the less it feels as if it is in our own hands. But the hard truth is that weight gain comes from eating too much, and it hurts to admit that at first. It's often easier for us, and those who love us, to soften that unpleasant blow with statements like, 'You've probably been eating too little — and now your body's gone into starvation mode.' (Which basically means, 'You haven't failed! Quite the opposite: you were *too good*!') It might be nice to hear such well-meant words of comfort at first, but in the longer term they just lead to frustration and desperation, because they effectively stand in the way of change.

Since I started actively paying attention to it, I've realised how rarely people who are frustrated by their failed attempts to lose weight are directly faced with the truth. Yesterday, for example, I came across a thread in a forum in which a woman was saying she ate very little and still hadn't lost any weight for months. These were the first four responses:

> If you consume far less than 1000 kcal per day for an extended period of time (and as far as I can tell that's the case with your eating habits), your metabolism will go into starvation mode. It's as simple as that. [58 upvotes]

> It sounds like your metabolism has reached zero point, which doesn't surprise me given the amount you're eating. [61 upvotes]

You're eating way too little. No wonder your metabolism's gone down the drain. [44 upvotes]

Maybe you're not eating enough? [...] Your body needs at least 1250 kcal to function properly.

Then came the obligatory comment about the possibility of an underactive thyroid. These are the sorts of responses I see almost every time someone describes this kind of problem.

Responses along the lines of 'you're eating too much' usually come in for hefty criticism and are dismissed as 'mean' or 'stereotyped'. In truth, the forum-poster who couldn't lose weight after months of 'reduced eating' was probably fooling herself and eating more than she thought. Persuading her that she needed to eat (even) more was the surest way to successful weight gain.

XY will speed up your metabolism

Unfortunately, metabolism seems to have become some kind of mythological altar to weight loss. People believe that if they could only find the right magical incantation and offer the right sacrifice, they might yet experience a miracle. 'Yeah, sure, calories and all that stuff, I know … But I heard if you drink a glass of apple cider vinegar before a meal, your metabolism won't absorb the fat.' Or, 'What about that Japanese medicinal pepper extract from the root of the seaweed tree? It was in the latest *Women's Weekly* — it speeds up your metabolism!'

'It speeds up your metabolism' has almost become an aphorism. When people say it, it always makes me wonder what kind of mental image they have of the metabolism. Is it a sort of engine somewhere below the stomach that turns fat into energy, where if it can be made to go fast enough, some of the food will shoot through it unprocessed? Or is it a little homunculus, shovelling food into a furnace, and if you take enough pepper extract, the little man will start shovelling so fast that all the food you eat will go up in a huge combustive flame without being used by the body?

If the people who say it knew that metabolism is actually our heartbeat, temperature regulation, and organ functions, would they still want to *speed it up*? If the heart is sped up so much that its racing wakes us up at night, that's not such a great thing, though it might burn a couple of extra calories.

If they really did discover some women's-magazine-supertrick to speed up the metabolism, then within days, doctors' waiting rooms would probably be full of people begging their GP to do something about their insomnia, tremors, hot flushes, restlessness, irritability, and racing heart. I'm sure no one would be jumping with joy and celebrating each pound they'd lost *so easily* that month.

The comparison with an apartment is fitting here: too little heating isn't nice, but no one can stand to live in an overheated flat for too long, either. There is a small temperature range that's pleasant, and of course, there are some small tricks to increase energy combustion a little (in other words, to heat the apartment just a bit more, from 18 to 20 degrees, for example). We should be clear here that statistically significant results might still be tiny. Even a difference of a few calories can be significant if the sample size is big enough. Most super-metabolism-tricks have very minor effects, and there's a natural limit to their possible effects anyway. It's not possible to speed up the metabolism to such an extent that it would really have a big effect on weight (by big, I still only mean about half a kilo per month, which most people wouldn't consider that much). And if it *was* possible, the side-effects would be so unpleasant and so dangerous that no one would want to do it.

There are certainly illegal dieting aids that work on this principle and that have caused deaths in the past. In April 2015, a 21-year-old woman died after taking diet pills containing the substance dinitrophenol. The drug really does increase the metabolic rate and cell activity, creating a lot of heat. The side effects include heart palpitations, cardiac arrhythmia, restlessness, overheating, nausea, and eventually organ failure. The *Daily Mail* put the young woman's fate into very vivid words, saying she had 'burned up from within'. And all this despite the fact that the substance was withdrawn as a dieting aid in the 1930s because the amount of weight loss it caused was so small.

More effective drugs usually work mainly by suppressing the appetite — Ritalin and other stimulants that are used to treat ADHD, for example, work this way. They stimulate patients to make them more able to function in daily life, but also suppress the appetite for the duration of their effectiveness. Once again, their effect is not due to a magically altered metabolism, but mainly thanks to reduced calorie intake (because of a lack of appetite).

This goes to show that the purely metabolic effect is extremely small in comparison with active energy regulation (through diet or exercising) and that there is no way a body can lose weight 'by itself' by simply increasing its resting energy requirement. Any suggestion by manufacturers in their advertising that this is the case is not just nonsensical, but also fraudulent.

What's also annoying about these kinds of metabolism-accelerating superweapons is that they mostly contain calories themselves, and often more than the number of calories they claim to *burn*. Apple cider vinegar also contains calories, and if the statistically significant effect of drinking it before a meal is five calories' difference, but there are four calories in the recommended 20 ml of vinegar, then anyone who forces the sour stuff down their neck isn't doing themselves much of a favour.

Many of these so-called silver-bullet remedies are sold as miracle weight-loss cures without even specifying precisely how their supposed effect works. Does it suppress the appetite, causing test subjects to lose 100 grams more per month by eating less? Does it work because of a faster metabolism? Most dieting aids don't work by increasing the amount of energy the body burns *per se*, but by making people move a little bit more in their daily lives, thanks to their slightly stimulatory effect — and that burns more calories. It's similar to the effects of caffeine or nicotine, which invigorate us, make us more physically active, and help us overcome bouts of tiredness.

One of the most effective dieting aids is green-tea extract, which has been shown to have a significant effect on energy expenditure, increasing it by around 4 per cent over a 24-hour period (Dulloo et al., 1999). For someone with an average daily expenditure of 2000 kcal, that means they would burn an additional 80 kcal per day. Not bad, but far less than necessary to lose half a kilo a month.

You might find all of this too negative and say, 'Cutting out an additional 80 kcal isn't bad! Why shouldn't I?' That is, of course, a valid argument. I can hardly talk up small changes on the one hand, and then turn around and say, 'Pah, 80 kcal of weight loss is pathetic!' But that isn't what I'm saying.

As I see it, the problem is mainly one of marketing and the suggestions it works with. It would be perfectly legitimate, in my opinion, if they said, 'Here's something that gives you energy, which will make it easier for you to be more active and burn more calories.' And if a would-be dieter responds with 'I'm willing to pay 20, 30, or 50 euros for that!' and is aware of the fact that the product can only be used as additional support, then it would only be a good thing. I drink relatively large amounts of coffee and green tea for similar reasons.

But if they hype a product as if it was a magical extract that can make you thin by mysteriously speeding up your metabolism, the result is that people will shy away from sensible strategies because they're more difficult, or they'll feel like fools for following those strategies when such an easy alternative is available. Not uncommonly, this is what the result of that looks like:

If you really want to increase your resting energy expenditure, there is basically only one useful adjustment dial you can turn up: muscle mass. Our muscles have to be in a constant state of readiness, requiring a certain basic amount of energy to be supplied by our body to maintain an electrical potential in them — unlike fat, which just hangs around. There are varying figures on exactly how many calories are burned per kilogram of muscle mass per day. On average, the various body-fat calculation programs reckon with around 30 kcal per kilo of muscle per day. So, if you achieve the ambitious target of gaining 10 kg of muscle mass by weight training, you will have earned yourself an extra half a bar of chocolate per day.

Gaining muscle mass by training is extremely laborious, though, and the typical initial progress at the gym is due less to muscle growth than to an improvement in the functioning of existing muscle mass as your body movements and coordination improve. A typical mistake is to assume that an increase in body weight after taking up training is thanks to 'muscle growth', rather than the chocolate bar you reward yourself with for every hour of sport you endure.

In my experience, a more realistic and practical way to burn more energy is to introduce little routines into your everyday life that mean you move more, thereby burning more calories. It's much more reliable than all the miracle cures, which can also pose a danger to health or even to life.

You're bound to put on weight
if you stop smoking

After spending the last few pages explaining that there is no such thing as a metabolic super-weapon, I now come to the topic of smoking. It's a widely accepted fact that quitting the habit leads to weight gain, and there are plenty of smokers who are afraid of quitting because it might make them fat.

The central question many people ask in this context is whether the weight gain is due to a metabolic effect, or to the fact that people who stop smoking are more likely to eat in order to compensate for the lack of oral stimulation they used to get from cigarettes. The answer seems to be both.

Just like caffeine or cocaine, for example, nicotine is a stimulant, and energy expenditure rises sharply by about 3 per cent after smoking. Several studies have shown that smoking's effect on the metabolism is equivalent to around 0 to 220 kcal per day on average, depending, of course, on how much a person smokes and how much they weigh — the increase in energy expenditure appears to be less for obese smokers (Hofstetter et al., 1986; Moffat & Owens, 1991). Interestingly, there seems to be a placebo component to this energy-expenditure-increasing effect. In a study (Perkins et al., 1989), increased energy expenditure in subjects who were given a placebo nose spray was still around half that of subjects who received the real nicotine nose spray. Researchers also believe that behavioural factors play a part in this.

This means that smokers who kick the habit should expect their energy expenditure to sink to a normal rate and fall by up to 200 kcal a day — the amount for which nicotine consumption has previously been responsible. Since nicotine also has an acute appetite-suppressing effect, new non-smokers are likely to feel more hungry. Stamford et al. (1986) found an increased daily caloric intake of 227 kcal among women who gave up smoking, leading to an average weight gain of around 8 kg within a year.

Of course, this isn't a valid argument in favour of smoking. Firstly, the habit can be just as deadly as the increase in metabolic rate described in the previous chapter. And secondly, no non-smoker would dream of thinking, *Ah, yes, I'm willing to pay quite a lot of money and accept a life expectancy thirteen years shorter, for the sake of up to 220 kcal a day.*

But even if someone *were* to find that tempting, there are also weight-related disadvantages to smoking. For example, smoking appears to increase the levels of cortisone and other hormones in the blood, and leads to a tendency for fat to be stored in the abdominal region. This increases smokers' risk of developing a particularly harmful type of fat deposit, and although their BMI may be lower, their weight-related risk of developing certain health conditions, e.g., diabetes, is higher (Chiolero et al., 2008). At the end of this book, I deal in more detail with the issue of people with a high body-fat ratio despite being of normal weight, but smoking seems to be a major factor in this phenomenon. Interestingly, despite the energy-increasing and stimulating effect of nicotine, smokers tend to be less active on average than non-smokers, and they tend to take less exercise. It remains unclear to what extent this is thanks to their general lack of interest in health issues or to their impaired physical capabilities as a result of reduced lung function.

There are many advice books and medical guidelines that claim that it's difficult, and therefore unreasonable, to try to lose weight and give up smoking at the same time. In fact, there's no scientific evidence for this. Spring et al. (2009) analysed studies on this topic and came to the conclusion that combining quitting cigarettes with a simultaneous diet does not produce worse results, and there may even be some evidence of the opposite. It seems some people find it easier to make dramatic changes to their lives all in one go and establish entirely new routines.

Another possible hypothesis is that skin regenerates better when it is not adversely affected by smoking. Post-operation wounds are known to heal better in non-smokers, and there is ample scientific proof that smoking accelerates the ageing process of the skin. So I can imagine that quitting smoking might very well have a positive effect on skin regeneration, which is another advantage when dealing with the temporary sagging skin during significant weight loss.

With regard to the fat logic in the title of this chapter, it only remains to point out again that giving up smoking does not automatically make you fat, just as pregnancy, illness, or ageing don't, as long as the right balance is struck between energy intake and expenditure. It might seem trivial to point it out again, but quitting cigarettes does not make the metabolism *slower*, but rather returns it to normal. So ex-smokers may find it subjectively more difficult to lose weight than if they still smoked, but that ultimately just means that their bodies have returned to the normal state shared by all their fellow humans.

Food with negative calories: eat more to weigh less!

On 20 February 2015, the *Daily Mail*, a newspaper with a readership of millions, published an article on foods which allegedly burn more calories than they contain. They were said to include cucumber, asparagus, cauliflower, celery, lean meat, tomatoes, papaya, chilli, and apples.

The internet is also awash with similar lists of foods that supposedly contain 'negative calories'. Those lists usually include fruits like apples, berries, oranges, and melons, but they may also contain vegetables, including asparagus, broccoli, courgettes, and cauliflower. Quite apart from the fact that none of these, nor any other food, contains fewer calories than they provide. It is doubly ironic that many of them, especially the listed fruits, are actually high in calories.

A 'calorie-free' apple comes in at around 80 kcal, and an orange has more than 100 kcal. And what about the 'calorie-free' melon? A small honeydew melon can easily contain more than 300 kcal. So, someone trying to lose weight by restricting their intake to 1200 kcal a day can easily eat their full allowance just by grazing on a few pieces of fruit through the day — or they can completely ruin their calorie deficit. Although vegetables are usually a little lower in calories, it's worth checking them on a nutrition chart. Carrots have between 20 and 40 kcal each, and if you're like me and often nibble away at a few carrots as a snack, you can easily take in 200 kcal that way.

The body does use a certain amount of energy in processing food, but only around 1 to 30 per cent of the energy contained in the food itself. For example, about half a calorie is required to digest 6 calories' worth of celery (Upton, 2008). That's only a fraction of what the body absorbs. Proteins normally require more energy to digest than carbohydrates or fat since the body finds it more difficult to break protein down and transform it into energy. So, a protein-rich diet isn't bad from that perspective, but it's still not a miracle cure that will allow you to eat unlimited amounts of food, or even lose weight by eating.

In the EU it's actually legally permissible to label foods 'calorie-free' if they have fewer than 5 kcal per 100 g. But even Coke Zero and Diet Coke have a tiny amount of caloric energy (around 1 kcal per 100 ml).

The only truly calorie-free — or rather, calorie-negative — foodstuff is water. It contains no calories whatsoever. But our system does need to use some energy to heat it up to body temperature and to process it. A couple of glasses of water require only a minimal amount of energy to process, but it is highly advisable to drink a lot of water while dieting anyway. As with everything, though, it's important to remember here that 'the dose makes the poison'. Suddenly downing several litres of water can create a mineral imbalance in the body, as it flushes out important salts. And that can cause serious health problems.

If you want to lose weight, eat more fruit

Even people who don't believe that fruit has 'negative calories' associate it with dieting. I have dedicated a chapter to this idea because it crops up so often.

As I have already said, food is never good or bad *per se*; it can, at best, be described as more appropriate or less appropriate for attaining a certain aim. For people who are losing weight because of illness or those who have a very high calorie requirement due to an active lifestyle, high-calorie foods like muesli bars can be a real blessing.

Whatever the case, fruit can't be considered 'slimming food'. A glance at the nutritional composition of various kinds of fruit is enough to see that most are mainly made up of sugar. Here, for example, is a table comparing the nutritional information for a large apple (200 g) to that of a glass of cola (approx. 200 ml):

	apple	cola
kcal	108	84
protein	0.6 g	0.2 g
carbohydrate	27.4 g	21.2 g
of which sugar	*26.6 g*	*21.2 g*
fat	0.4 g	0 g
fibre	4.8 g	0 g
water content	80 %	90 %

A large apple contains more sugar than a glass of cola, and the rest of the nutritional content (carbohydrates, fat, protein, water) is pretty much the same. The only major difference is in the dietary fibre content, which is much higher for the apple. Dietary fibre slows the absorption of sugar, making the apple a gradual and steady source of energy, whereas the cola causes a sudden spike in sugar levels. That's why the apple leaves you feeling fuller for longer. The apple also contains considerably more vitamins. So, drinking a glass of coke and taking a vitamin pill is not radically different from eating an apple, as far as the nutrients are concerned. With its high sugar content, fruit should be seen as a dessert or a sweet snack for between meals, but it should never be seen as something to eat *in order* to lose weight. In other words, fruit is a healthier alternative to other sweet treats. Replacing a bar of chocolate or a piece of cake with a piece of fruit will usually save a few calories (e.g., a large apple, 110 kcal; a small chocolate bar, 130 kcal). It will also make you feel fuller for longer.

Unfortunately, a lot of people wrongly consider eating fruit to be an active way of losing weight. That means, rather than being a sweet dessert, eating an apple becomes a healthy chore. Some dieters stuff themselves with masses of fruit in the belief that it will help them lose weight, and feel they have been 'good' if they polish off three apples as a between-meal snack, whereas in fact they have consumed more calories than if they had been 'very naughty' and treated themselves to a bar of chocolate.

So for those who can enjoy fruit as a tasty alternative to sweets: good for you! But people who are not particularly keen on fruit should not be made to believe they have to force down mountains of it to lose weight. If that's the case, it makes more sense to have some chocolate, and give yourself a treat that is really worth the calories. After all, although fruit leaves us feeling fuller for longer on a purely physical level, the psychological satiety provided by *naughty* foodstuffs is a factor that should also be taken into consideration.

Slimming products just make you fat

I'm sure some readers will be asking themselves what this chapter's title has to do with fat logic. Everyone knows that sweeteners and diet products actually make us totally fat — after all, sweeteners are used to fatten pigs, and 'diet' or 'light' are just advertising slogans and a big con.

On the other hand, careful readers will have noticed that I espouse the caloric-balance approach. This dieting method doesn't need to be proven by masses of scientific studies; a glance at our supermarket shelves is enough. Light and diet foods often really are lower in either sugar or fat. Sugar is usually reduced in diabetic products, fat in calorie-reduced foods. Since both are flavour carriers, one is often replaced by the other. Products that purport to be fat-reduced contain more sugar, and low-sugar products have more fat in them. Sugar-free chocolate is often as high or even higher in calories than *normal* chocolate. But that's not always the case. Some slimming products really do contain considerably fewer calories; there is no general rule. As with all products, the only way to be sure is to read the actual nutritional information given on the packaging. Personally, I have nothing against diet products and relatively regularly use some of the ones that do provide a good alternative.

For a long time, I was one of the people who spread the myth that artificial sweeteners cause the body to release more insulin, leading to a fall in blood sugar levels and therefore to cravings and snack attacks. In fact, this effect has never been confirmed, and studies on the link between sweeteners and weight loss have yielded ambiguous results. I found some studies with positive results, and others whose conclusions are negative, so it seems impossible to say with any certainty whether sweeteners work more to suppress or to stimulate the appetite. An overview analysis by Bellisle & Drewnowski (2007) concluded that 68 per cent of studies on this topic found there to be no effect; 16 per cent saw a negative effect, and another 16 per cent found the effect of sweeteners to be positive.

However, the myth that sweeteners are used in the farming industry to fatten pigs has been disproven several times — in the German newspaper, *Die Zeit*, for example, in which a 2008 article explained that artificial sweeteners are only used to help wean piglets off the sweet milk they receive from their mothers and help them get used to their new food.

I suspect that the effect of artificial sweeteners is more of a psychological one. Some people will find it easier to just ban all powerful sweeteners from their diets and develop a finer taste for the natural sweetness of fruit, for example. But others prefer the sweet alternative of a Coke Zero, leaving them with less of a feeling of having denied themselves a pleasure.

Personally, I like the taste of some sweeteners very much, and I enjoy Coke Zero and other diet drinks. Unfortunately, stevia doesn't really agree with me, but I like to use other sweeteners in my coffee, yoghurt, or desserts. I sometimes come in for criticism for this, for 'cheating' or using a 'cheap alternative'.

I find that reaction plain silly. I don't see why I should have to consume extra calories out of pure principle, especially when I don't experience any discernible difference in taste. I'm certainly willing to believe that some people don't like the taste of some artificial sweeteners. Of course, it makes no sense to use these products if you don't like the way they taste, just so you can eat twice as much of the artificially sweetened food than otherwise. But the almost fanatical idea of 'avoiding diet products at all cost' is just not founded in fact. If you like the taste, there's no reason not to eat it. Especially given that the supposed negative effects on health and carcinogenic properties of artificial sweeteners have remained completely unproven by decades of scientific study. Germany's Federal Institute for Risk Assessment has not even found it necessary to issue a maximum safe limit for sugar substitutes such as sorbitol, erythritol, and xylitol.

At this point, I had intended to describe briefly the effect of sugar on the body and compare it to that of artificial sweeteners. However, my researched revealed that the numerous studies on sugar and its health effects are extremely inconsistent in their results. This is partly because many of these studies rely on self-reporting by subjects, rather than objective measurement. That means that false perceptions can heavily skew their results. Many of the observational studies also fail to clarify whether the negative effects observed in connection with high levels of sugar consumption are actually due to the sugar itself, or to do with the concomitant increased caloric intake and weight gain. Studies based on experiment have also failed to come up with unambiguous results. Black et al. (2006), for example, found no ill effects of a high-sugar diet on insulin resistance, blood sugar levels, or the vascular system in healthy test subjects, although there was a slight increase in the subjects' cholesterol levels. But Black et al. do make reference to an earlier, smaller study which observed clearly negative effects.

It seems this plethora of contradictory research makes it difficult for experts to give any well-founded statements about sugar. In response to a question from me, the nutritional epidemiologist Dr Gunter Kuhnle, who carries out research into sugar, expressed scepticism about the demonisation of sugar in the media. He believes it is due to the connection between high levels of sugar consumption and increased calorie intake and resultant obesity. He is of the opinion that sugar has a place in a healthy and balanced diet as long as people don't overdo it. He also believes that the WHO recommendation that no more than 5 per cent of a person's total caloric intake should come from sugar is a reasonable one.

It's difficult to identify any clear position in the general confusion about whether such-and-such a sweetener is bad, very bad, or just a little bad.

Returning to my main point, what remains clear is that more calories lead to weight gain and fewer to weight loss. So, when slimming products form part of a calorie-reduced diet, weight loss will result in just the same way as it would with the full-fat versions — as long as the number of calories consumed remains below the set limit. Once again, the only danger here is the self-deception these products can lead to: anyone who likes to eat sugar-free chocolate, which actually contains more fat and therefore has 50 kcal more than ordinary chocolate, and then reasons, 'It's diet chocolate, so I can eat as much as I like', will, of course, gain weight. This isn't because slimming products make you fat, but because overeating makes you fat.

I've already said it several times both directly and indirectly, but it cannot be repeated often enough: no specific kind of food makes you fat or makes you thin. Some foods might make gaining weight more likely than others, but dividing food into good and bad is nonsense. If I am exercising intensively, for instance, and need a quick source of energy to get me through a few more kilometres, a sugary drink is the perfect thing. There is ample evidence that simple carbohydrates increase athletes' performance during long endurance sessions (Jeukendrup, 2011; Wright et al. 1991). A glass of apple juice or Fanta isn't really necessary during normal daily life because there's usually no need to rapidly increase your blood sugar level in the office at work. But during physical exercise, that surge of energy is consumed immediately, so it makes sense.

Likewise, it's wrong to reject food that's high in calories as automatically unhealthy. High-calorie foods like nuts, olive oil, or salmon, for example, contain healthy fats, and aren't 'bad' just because they are densely packed with energy. They are perfect for people who want to gain weight, and for people who are trying to lose weight, but who like those particular foods, they can also be integrated into a weight-loss nutrition plan.

Calorie counting is an eating disorder and spoils your enjoyment of eating

In this book I talk a lot about calories, and some readers may have formed the impression that I think the only way to lose weight is by counting calories. That's not the case. Counting calories works for me, but it is just one tool among many.

I have never personally been told that calorie-counting is an eating disorder, but I have read it many times in articles, blogs, and forums. The *Daily Mail*, for example, published an article in February 2015 claiming that people who are 'obsessed with counting calories' skip meals during the day to save up their calorie allowance, which they then use to drink in the evening. The paper labelled this behaviour 'drunkorexia'.

According to this logic, we would also have to say that knives and hammers are evil because they can be used to murder people. There is no doubt that it's possible that the symptoms of someone with an eating disorder might include calorie counting, but that doesn't mean counting calories is necessarily a sign of a disorder. And I can't see any logic in saying that counting calories in food is okay, but those in alcohol should not be counted because *that* is the sign of a disorder. If you only count some calories, but not others, you may as well not bother in the first place. The journalists who wrote that article appear to be of the opinion that the body turns a blind eye, or a blind gut, to the calories in alcohol. By that logic, you could also say 'If I don't list the twenty euros I spent on cinema tickets in the household accounts, then the money won't have gone from my wallet.'

I think we have to start admitting that intuitive eating doesn't work especially well for more than half of the overweight population in our society. Weight is increasing in general. And if many people didn't counteract that with dieting, exercise, and other weight control methods, the number of overweight and obese people in the population would probably be even higher. There are those who intuitively eat the right amount and engage in exercise throughout their lives, but such people are clearly not in the majority.

People who can eat well intuitively and have no nutritional deficiencies, good blood test results, and a healthy BMI, can continue as they are wonderfully, of course, without any outside influence. The large group to whom that does not apply, however, including the overweight, the underweight, the obese, or those with unhealthy blood test results, will at some time reach a point where they realise that it would be sensible to take a good look at their eating habits.

In my opinion, counting calories is a good way to approach this, as modern technologies (calorie-counting apps) can show all the nutrients included in a given foodstuff. It is not difficult to look up different foods and put together a diet plan aimed at achieving a specific goal.

After a friend of mine started counting calories, she told me how surprised she was at the number of calories in some foods that she had considered healthy (unfortunately, healthy is often used as a synonym for low-calorie): nuts, for example, or fruit. Other food, like low-fat mozzarella — which she preferred — was much lower in calories than she'd thought, and was a good addition to her diet plan, helping her achieve her aim of eating more protein.

I can understand that some people don't like the mathematical nature of this approach and that they'd prefer to lose weight using a different strategy. My mother started trying to lose weight soon after I did, and her approach boiled down to 'move more, eat less' (and 'no carbs in the evening', although the time of day is basically unimportant). She managed to go from an obese-level BMI to one that's only slightly overweight, losing around 30 kg overall.

It is possible to lose weight without counting calories, but I do find the claim that calorie-counting is somehow a sign of an eating disorder pretty outrageous.

For a long time, I worked in a full-time job and lived in rented accommodation. During that period, I always had enough money in the bank and never really had to think about it. I don't think I would have looked at my balance more than once a month, and I never spent time worrying about the price of things while shopping (I'm naturally quite a thrifty person). Now, we've bought an old house that needs a lot of work, and I am self-employed. That means I have to keep a keen eye on my bank balance whenever I am planning to spend money. Does that mean I have a 'financial disorder'? Have I destroyed my 'intuitive spending' by obsessively checking my account?

The same holds true for other situations where a lifestyle change means 'intuitive' no longer works: when you have a physically demanding job and you're suddenly transferred to the office, and you begin to notice your trousers are getting tighter at the waistband; or when you start a new relationship and you and your partner enjoy many evenings in, cooking together; or when injury means you can't exercise for a while. A sudden change to a daily routine means a change in caloric balance, which some people are to some extent able consciously to control. But sometimes it's necessary to take the initiative and actively examine your caloric balance to avoid running up any *debts*.

Incidentally, counting calories is not just a useful tool for overweight people. As part of my research for this book, I took a look at some of the anorexia forums online and spoke to a friend of mine who works in an assisted-living project for people with eating disorders. It's not uncommon among these patients to have an extreme phobia of certain foods, because they are thought to somehow 'instantly' make you fat. Calories are a constant issue for people with anorexia, but their concerns are often unfounded in real fact. I'm certain that some people with eating disorders would benefit from accurate calorie counting as a way to help them overcome their irrational fears. As I see it, the thought processes that so entrench the underweight state in the minds of anorexics are often very similar to those that prevent overweight people from slimming down: controlling your body weight undergoes a kind of negative mystification and becomes something difficult and highly complex, which has more to do with beliefs than reality.

What I found so helpful was the fact that by keeping an overview of my calorie intake and thereby uncoupling it from fluctuations due to fluid retention, I had a good idea of how my weight loss was really progressing. That meant I could easily take it in my stride when I had a daily deficit of 1000 kcal and still saw an extra pound on the scales the next day. I knew for certain that the increase was only due to water, and that I had actually burned off 140 grams of fat. And if the scales showed a gain of 2 kg the day after a big holiday meal, I could rest assured that I hadn't eaten an extra 16,000 kcal, so most of that short-term gain must not be fat.

By contrast, someone with an eating disorder would be likely to panic at a sudden gain of 2 kg after a holiday blow-out, with an instant disaster scenario in their minds: 'If I keep going like this, I'll have gained thirty kilos by the end of next week!' Which is, of course, totally unrealistic.

A realistic understanding of calories would be helpful in these cases, as it would also come with the knowledge that no one can suddenly gain 30 kg from eating a little more than normal over the holidays, and nor can you have lost a kilo by eating nothing but chocolate, just because the day after the chocolate binge also happens to be the day your body retained less fluid.

The day's ruined now, anyway

This is by far the stupidest piece of fat logic of all, but one that I applied passionately in the past. And, unfortunately for me, it's also the one that I find hardest to banish from my mind. Counting calories and understanding fluid retention helped me to realise how completely irrational this way of thinking is, but it's very difficult to throw off. At least, it is for me.

I still sometimes catch myself thinking that that second helping doesn't matter now, anyway, or that since I've already overdone it today, a bar of chocolate won't make any difference.

The little brother of that kind of fat logic is this:

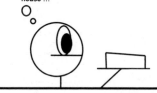

Some residual fat logic is probably always going to be present, and overcoming it is a constant work in progress. It's the same with that packet of biscuits you really should polish off before … oh.

Of course, such extreme thinking can put you in a situation where you feel like you've regressed into old patterns of behaviour, that you've 'slipped', and that you've lost your motivation. Over the past ten years I've started numerous diets, only to pack them in after a few days or weeks, mostly when I stopped seeing any results. But this time, the knowledge I'd gained about the mechanisms and processes in the body helped me to avoid falling into that classic diet trap. I kept it up for nearly six months without any significant problems. Then I had my knee operation, and the doctors told me that I definitely ought to put my diet on hold and start eating more, so that my body could regenerate. I don't think they meant that I needed to order pizza the evening after the operation; that was my own interpretation.

Huge amounts of fluid retention, and a guilty conscience at having totally overshot the mark, were the unpleasant consequences of that excess. But the fact that I quickly got my act back together turned out to be important for my future progress. That 'slip' and my subsequent ability to 'get back on track' quickly were an important development for me because that's when I realised that I'd changed my way of thinking.

The experience also helped me when I faced my next big test — which I initially flunked in grand style. When I started a two-week holiday in September 2014, I had just about reached normal weight. The plan was to spend a week visiting my parents, and then a week staying with my mother-in-law. Even as I was planning the holiday, I knew there was little chance of losing weight during that time. But I had resolved to try to keep my weight stable, or at least not to gain too much. At the end of my fortnight away, the scales showed seven extra kilos. A good part of that was retained fluid, but some of it was also fat.

The week I spent at my parents' home, in particular, was a combination of everything that made sticking to my regimen difficult: the entire house was full of my favourite foods, especially sweets and my grandmother's baking; there were lots of family parties where we ate out; my mother spoiled us with huge breakfasts; and there always seemed to be coffee and cake wherever we went. There were many days that I wrote off completely, and after just half a week I knew I'd finish the holiday several kilos heavier than when I started it. But what I'm particularly proud of when I look back on that vacation is that I still didn't go to the absolute extreme, as I would have before. I refused to overindulge on calories on several occasions, did some exercise, and even managed to have a calorie deficit on some days. Although the tendency was bad, I didn't slip completely, and that was an important experience for me.

And more importantly, I remembered back to my first slip and how I'd managed to stick to my diet plan nonetheless. That helped me overcome my feelings of resignation and reluctance. Even though it took almost a month to get back down to my pre-vacation weight, I stuck at it.

I am now convinced that those two experiences were among the most important in my weight-loss process. The more I increased my caloric intake in the subsequent months, the more slips I made. Remembering the fact that I'd already coped with two major slips and had still managed to get 'back on track' was a big help in that time.

I was also helped by something I'd read in an online forum: 'It's not about single days, what's important is the big picture.' Like others, I got fat despite individual diets and single days spent eating not much. Most overweight people briefly go on diets every so often. But despite these efforts, I still weighed in at 150 kg. My weight gain had not been hindered by the several weeks I spent on diets every now and then. But the opposite is also true: single slip-ups, or even bad weeks, don't change the general trajectory. As long as you always return to that basic direction, those slips don't make an important difference. It's not important to eat a perfect diet 100 per cent of the time; what is important is to achieve a calorie deficit in balance over the long term.

For me, this all-or-nothing way of thinking is the most persistent and difficult aspect of fat logic. Paradoxically, it was my slip-ups that turned out to be a help in the struggle against this kind of fat logic. I'm sure that if I'd lost weight over a year in a perfect and easy process, I would now find it much more difficult to maintain that weight. There are many more grey areas when it comes to maintaining your weight, and there are days when I eat a little more than my daily requirement, and some when I eat a little less. Now, if you haven't already, is the time to put aside that extreme way of thinking. Doing it will mean that after a few days of cravings and slips, you'll be able to gently turn yourself back in the direction of a little less, rather than thinking, 'Oh, it doesn't matter now, anyway,' as the sales assistant wraps up your doughnuts at the bakery.

You should never lose more than a pound a week

This is the myth that I encounter most often. The reason for its stubborn persistence is probably because it's repeated so often by so many different people, including doctors, dietitians, and physical trainers, not to mention all the everyday weight-loss 'experts'. The explanations given for this rule vary a lot, but none of them are based on fact.

'You'll be malnourished, and you'll ruin your health.'
Of course, the more calories you consume, the more likely you are to get enough vitamins, protein, and minerals. But someone who eats 3000 kcal in the form of fries, chocolate, and pastries is certain to get fewer important nutrients than someone who takes vitamin supplements and has a small, protein-based meal of 600 kcal. According to the *National Post*, people who overeat are more likely to have an unbalanced diet and to be deficient in certain nutrients. It seems paradoxical: few people would imagine that an overweight person could suffer from malnutrition, but it's often the case. And I was a case in point: at the start of my weight-loss process, blood tests showed that I was suffering from iron, vitamin-D3, and vitamin-E deficiency. My blood pressure was also far too high, I had problems sleeping, and I suffered from exhaustion and constant pain.

I was able to redress all those imbalances during the first six months, when I had a daily caloric intake of only 500 kcal. My blood pressure went down to a normal level and my physical fitness improved just from losing weight, given that I wasn't able to exercise. During that period, I was losing an average of 7 to 8 kg per month — four times the supposed healthy weight-loss limit. This shows that eating a lot doesn't necessarily mean eating the right nutrients, and eating little doesn't automatically mean malnourishment.

One downside was that my uric acid levels increased while I was losing weight, which raised the risk of an attack of gout. My doctor explained that this effect is possible when the body breaks down a lot of fat. With regular blood tests, reducing my uric acid levels using medication was not a big problem. And I was able to stop taking that medication after I lost weight, with no long-term damage done. Losing weight quickly is not automatically bad for your health.

On the other hand, being overweight is demonstrably damaging for your health. If I'd followed the advice not to lose more than one pound a week, then instead of being within the normal, healthy weight range after a year, I would still have been within the range of morbidly obese. That recommendation is plainly unrealistic for severely overweight people because many weight-loss strategies — like sport — aren't practical for them until after they have already lost massive amounts of weight. Exercising with that kind of weight puts a lot of strain on the body, so it can make sense to lose a bit of weight first before putting more strain on an already over-strained body.

The situation may well be different for someone who just wants to lose a couple of kilos and is either of normal weight or only slightly overweight. In a case like that, it might be sensible to introduce a small deficit and adopt it as a realistic, long-term lifestyle change for the future. When my BMI reached about 24, I also shifted my strategy — towards slower weight loss with a focus on exercise.

I ate quite a lot, but I also did a lot of exercise, and I was only losing around 2 to 3 kilos a month. I don't think it would have made sense to lose weight that way right from the start and so to force my body to spend more months massively obese, *and* put it under extra strain from exercising. I don't imagine it would have been the death of me, but it would have exposed my body to more wear and tear and forced my joints and heart to work extra hard.

'Losing weight too quickly ruins your metabolism.'
I've already dealt with this at length in the chapters about metabolism. You can't ruin your metabolism!

'If you lose weight quickly, you'll put it back on quickly, too. If you want to lose weight permanently, you have to do it slowly.'
All the scientific papers I have read on the topic of maintaining weight loss reach the same conclusion: faster, greater weight reduction leads to better results in the long term. The probability of maintaining your lower weight increases, if you, (1) have lost it with a very heavily calorie-reduced diet, and (2) the amount of weight lost was large. Anderson et al. (2001), for example, concluded that the

amount of maintained weight loss was significantly larger after a diet of fewer than 500 kcal, or sometimes fewer than 800 kcal per day, or when the amount lost was more than 20 kg, than with diets that resulted in slower weight reduction or an eventual weight loss of less than 10 kg.

The study published by Thomas et al. (2014) also shows that a larger initial weight loss was associated with better long-term results after ten years. Purcell et al. (2014) found no difference in the long-term success rate of a group who ate between 450 and 800 kcal per day for three months and one that lost weight over a period of almost nine months with a daily caloric deficit of 500 kcal (which is equivalent to a weight loss of about one pound a week).

The myth that losing weight rapidly means you will regain it rapidly persists, although the opposite appears to be true in reality.

Personally, I don't think there is one general rule for everyone. Some people might feel better and more successful when they lose weight slowly but steadily. Others benefit from rapid results. I believe the psychological effect is paramount here. On a physical level, it makes no difference whether we empty our fat stores slowly or quickly.

Our physical bodies are not separately sentient beings that 'get scared' when we lose weight quickly. But our emotions do play a part in the process, of course. For some people, rapid change is frightening and difficult to deal with. Others, like me, need to see success quickly to remain motivated. In my case, the effect of rapid change was that I found it easier to compare before and after. The transition wasn't gradual and unnoticeable, as it had been with my weight gain, which progressed at about a kilo per month, leading to a gradual habituation process and a failure to realise that my fitness level and my health were gradually worsening all the time.

While losing 7 kg per month, I was able to look back to the previous month and see a distinct change. I could see it in the mirror, but also, more importantly for me, I could really feel it, too. I experienced directly how weighing 7 kg less made it easier for me to climb stairs, and I could feel how much easier it was covering the same distance in March than it had been in February. I was now able to lie on my back again without feeling as if my upper body fat was pressing on my throat. It was enormously motivating for me to physically experience those unmistakable differences. It meant I could clearly envisage the even greater improvements that were possible in the following month, or in three months' time.

If I'd forced myself to stick to the 'only one pound a week' rule, I would not have felt those effects. And I'm sure that, as a result, it would have been much harder for me to stick to my guns.

In this context, it's good to examine your own psychological makeup: are you the impatient type, who needs to see results quickly and is willing to invest in achieving them; or are you more the type who prefers discrete, gradual changes so that you approach your goal more slowly, but without feeling that you've had to deny yourself too much? Of course, it's possible that it might make sense to consciously choose one path or the other depending on your physical circumstances. Those who have medical conditions and need to eat certain foods might not be able to change their diet very much without negatively affecting their health, and so have no choice but to lose weight slowly. On the other hand, for me, it was important to remove the strain from my damaged knee as quickly as possible so that 80 kg of additional weight didn't impose on the already injured bone for any longer than necessary.

'Losing weight too quickly causes eating disorders.'

The previously mentioned Minnesota Starvation Experiment is often quoted as a warning against losing weight too rapidly by means of extreme caloric reduction, as it can lead to eating disorders, especially binge eating. The explanation put forward for this is that it basically inevitably causes uncontrollable urges to eat.

But is that really the case?

A recent study carried out by da Luz et al. in 2015 investigated this question by analysing previous studies on extreme caloric restriction. Their findings were mixed, but overall they discovered no proof that people who had no previous history of eating disorders developed them due to very low caloric intake. Among those who had been diagnosed with a binge-eating disorder, it was even found that extreme caloric restriction was able to reduce their symptoms. This led the authors to conclude that a medically supervised, extremely calorie-reduced diet can be helpful in treating binge-eating disorders.

My personal experience is similar. I was not a binge eater, but during the six months in which my calorie intake was restricted to 500 kcal a day, my appetite became very much smaller. Unlike previously, when I'd been able to eat at any time, those 500 kcal began to seem like a very large amount, and every incremental increase in my caloric intake felt like I was suddenly eating massive amounts of food. Unfortunately, the effect didn't last forever. I remember how, when I'd reached about 1000 kcal a day, I thought, 'Good heavens. Later on, when I'm just trying to maintain my weight and I'm eating twice this number of calories, it'll be an enormous amount.' Now, on exercise days, I eat far more than 3000 kcal, and it doesn't feel like such an enormous amount. However, the time I spent on 500 kcal per day worked like a sort of 'reset', subjectively at least.

I think we quickly get used to eating a certain amount, and we can have very different perceptions of changes to that quantity. If I'd reduced my intake gradually from my initial 3000 kcal a day, each of those incremental reductions would have felt like a permanent restriction. The effect of reducing my calories so drastically was that I experienced each increase as a huge luxury. So, once again, there is no general rule for how people will react psychologically to more or less drastic caloric reduction. For some people, the result could really be to stimulate their appetite, others might see a positive change in their relationship to food or to their perception of portion size.

'Losing weight too quickly means your skin won't be able to get taut again.'

As yet, there is little empirical research into this topic, but there appears to be no significant connection between the rate of weight loss and the skin's ability to regain its tautness. Your skin also doesn't stop tightening at the moment you reach your target weight. For months after, it continues to contract — in fact, it is easier for the skin to tighten at that stage, because it's no longer being stretched by fat. In many cases, the sagging 'apron' of abdominal skin so often seen in people who have recently lost weight is not actually just made up of skin, but also fat. You can understand this easily if you pinch the folds and feel the mass they still contain. Pure skin is paper thin. In these circumstances, the skin is unable to shrink because it's pulled downwards by the fatty tissue and gravity. The body needs to lose more fat before the skin has a chance to become tighter. Unfortunately, many people stop losing weight once they have reached a normal level, but they still have relatively large amounts of body fat and, if they don't do weight training, very little muscle mass. Since the skin and remaining fat tend to be saggy, the visual result is rather ordinary, and sagging skin can also cause health problems, such as infections between the folds.

If for no other reason, the statement in the heading of this section seems illogical to me because fatty tissue has a naturally limp structure and so does not provide the skin with any support at all. Skin may be 'filled in' by fat, but that can't be construed as 'tightness'. The best way to make sure skin is as tight as possible is not to lose weight deliberately slowly, but to build muscle mass to provide general tissue support and a replacement 'filling' for the skin. It's also important to leave the body with as little fatty tissue as possible to drag the skin down. An overstretched rubber band doesn't regain its shape any better when it is slackened *really* slowly. The best way to let it regain its shape is to stop stretching it at all.

For those who are just overweight, rather than obese, sagging skin should not be a major concern. But those who have always weighed more than twice the normal amount should try to keep their expectations reasonable in this regard. It's a bit unrealistic to expect a body's skin to contract back to normal size within a year, after it has been stretched to more than twice its natural size.

When I published an article in 2015 that included an image of my belly, I got a lot of responses, including a few questions about my 'rolls of skin'. My answer was that they were not skin, but 'residual fat', and that I intended to remedy the situation by losing more fat and training my muscles for the skin to 'lie on'. At the time, I wasn't completely sure whether I was correct in my interpretation. But I turned out to be right: now, at the time of writing, it's mid-March, so a month after I published the article, and this is how my 'rolls' had developed:

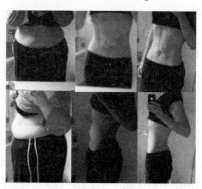

The first two pictures on the left were taken in December 2013, when I weighed 145 kg; the middle photos (weight:

65 kg) are from February 2015; and the images on the right (weight: 63 kg) are from March 2015. As you can see, my skin eventually really did 'lie on' my muscles as soon as it stopped being stretched by any residual fat. This might not work for extreme amounts of sagging abdominal skin. But it's clear that slightly to moderately saggy skin benefits more from losing additional fat and building up muscle than from slow weight reduction. From personal experience, therefore, I would say that the main way to get tight skin is a thorough process of fat reduction.

I assume that the reason losing weight slowly is thought to produce better results in terms of sagging skin is the longer comparison period. When someone loses weight quickly and makes a before-and-after comparison within a year, their skin will probably look slacker than that of someone who loses weight slowly and makes the before-and-after comparison after three years. But skin doesn't stop getting tighter at the moment when that final kilo is lost. If the fast weight-loser makes another before-and-after comparison after three years, his or her skin will probably look much tighter.

There are several things that can help you to gain tighter skin after losing weight:

- weight training, to fill some of the lost 'mass' with muscle;
- endurance training, to stimulate the blood supply to the skin;
- a vitamin-rich, high protein diet, to promote the repair process;

- moisturiser, to stop your skin from becoming too dry;
- and cling wrap and coffee-ground body wraps — as weight-loss methods, these kinds of poultices are quackery, but they can actually help remove fluid from body tissue and so help to make your skin tauter. Wraps don't reduce fat, but they do reduce girth and promote tighter skin.

I could never be really thin, my body isn't built that way

One reaction I often get from people who see me now is astonishment. Lots of them say they never would have believed I could ever be this slim. I often used to encounter this way of thinking among overweight people — people who had been overweight their entire lives and were absolutely convinced that they were unable to achieve normal weight because of their 'big bones' or 'heavy build'.

One example of this is an American blogger weighing 180 kg who wrote an article publishing the results of her body-fat measurement, which supposedly showed that her 'fat-free mass' would alone indicate a BMI of more than 25, proving that she was incapable of achieving normal body weight. But that calculation is based on the assumption that a body's fat-free mass doesn't change. The fact is that muscles and bones have to adapt to carrying three times their normal weight and therefore weigh more themselves than they would after weight loss. That means that fat-free mass reduces along with body weight. If you just piled 100 kg of fat onto a woman weighing 60 kg, she'd collapse. Her bones would have to grow denser and her muscles would have to become considerably stronger in order to carry that extra weight. If you were to suddenly remove all the excess fat from the body of the abovementioned blogger, it would reveal the figure of a body builder. It follows logically that such overdeveloped muscle mass is no longer necessary when the 120 kg of extra fat are no longer there, and so her musculature and skeleton would return to a normal size.

After I lost weight, my ribcage stuck out excessively for several months. My physiotherapist suspected that it might be because of my enlarged organs and the fat in my chest and abdominal areas, and that it would recede, and she turned out to be right. Initially, there was a massive discrepancy in my clothes sizes: I could soon fit into size 36 trousers, but I had to buy tops in size 40. I now wear size 36 for both, as my ribcage is now significantly smaller. See the comparison here:

 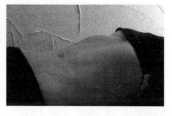

The picture on the left was taken in November 2014, when my weight was 71 kg. The photo on the right is from March 2015, when I weighed 63 kg. My ribcage sticks out much less in the second picture. I think that this was the reason, too, that people in the upper ranges of normal body weight and lower ranges of overweight commented that I was already so 'skinny': my expanded ribcage made my belly look 'hollow' and my lower abdomen looked smaller than it really was. Now, with my body parts in proportion and the overall impression more harmonious, I'm hardly ever told that I'm 'too skinny'. Comparing these two pictures also really brought home the health aspects to me again. It's almost frightening to think about the amount of fat that must have been in and around my organs to push my ribs out so far.

In general, it's nonsense to use having a 'heavy build' as an argument for being unable to achieve normal body weight. For me, normal body weight ranges from 56 to 76 kg. That 20 kg difference is enormous and certainly enough to cover both a 'slight' and a 'heavy' build.

My feeling is that this 'impossibility' is really psychological in nature, and that many people, especially the ones who have been overweight for years, have adopted a 'fat identity' to such an extent that, although they may be able to imagine being 'less fat', the idea of their ever being 'thin', 'petite', 'athletic', or similar is completely beyond the scope of their self-image. A lot of overweight people I know who decide to try to slim down set themselves a target weight that's still within the overweight range because a normal weight would be 'too extreme' — they don't want to end up 'all skin and bones', after all. A lot of people found it odd that I set my target weight at around 63 kg, with a height of 175 cm, when 75 kg would have been 'enough'. For a long time previously, I'd said that, ideally, I would like to weigh what I'd weighed when I'd met my husband — about 85 kg — because I couldn't imagine having a permanently slim figure.

In fact, the difference in calorie requirement isn't that big: with a weight of 85 kg, my body would need around 2000 kcal a day; with a weight of 63 kg, my daily requirement is a little below 1800 kcal. But the big difference is that with a weight of 63 kg, it's much easier for me to exercise, and keeping physically active is more enjoyable. This additional activity raises my daily requirement to more than 2000 kcal, and so, paradoxical as it might sound, it's therefore easier for me to maintain a target weight of 63 kg than 85 kg.

I'm sure there are people who feel more comfortable being 'a little chubby', for whom that decision makes sense and isn't associated with too many health risks. But I get the impression that a lot of people impose limits on themselves or have limits imposed on them by their environment, because they're clinging on to their 'fat identity' and believe that they are naturally unable to become slim. In bad cases, this belief can mean that you wind up with sagging skin and that your physical fitness never gets as good as it could be. You might be 'less fat', but the change isn't that significant. To me, this belief actually makes it much harder to maintain your target weight, because you imagine you're forcing your body into a weight range in which it doesn't really belong — because its natural state is more 'sturdy'.

It might sound strange, but I think it's important to realise that anyone can theoretically achieve and maintain any weight. A normal BMI has a large range within it, from 'very slim' to 'hefty', and your final goal should be dictated only by your own taste or wishes, and not by any ideas about how your body is 'naturally' constructed. Genuine 'body types' refer to nuances of bone structure — slightly broader shoulders, narrower hips, knock-knees, or bowlegs — those are the real physical differences between us.

The fat that lies on top of all this is determined by behaviour alone, and 'fat' and 'thin' are not 'different body types'. Many of my physical 'characteristics' didn't show themselves, or even began to change, as my fat gradually disappeared. For example, I thought I was naturally knock-kneed, but in actual fact, it was just my fat inner thighs pushing my knees together and making my legs look as if they were X-shaped. These days, my legs are straight.

[Male fat logic] Being a normal weight would make me a 'puny weakling'

Like the blogger in the previous chapter, some overweight men I've spoken to argue that their muscle mass alone would make them 'slightly overweight' and so a body weight within the normal range would leave them looking far too skinny. But these men are rarely highly physically trained body builders with low body-fat values, but rather men who are carrying a fair few spare tyres on top of those muscles. While it's extremely unlikely for a woman to be overweight due to muscle mass, it's definitely possible for men — but with a probability of 3 per cent, it's not exactly the norm. Three Reddit posters have been kind enough to let me to use their photos to illustrate what a man with a normal BMI, who works out, looks like:

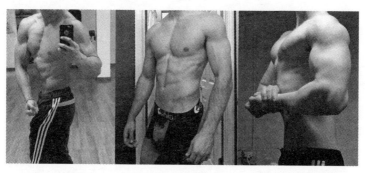

173 cm, 73 kg (BMI 24.4)
www.reddit.com/user/50_shades_of_gains

183 cm, 82.5 kg (BMI 24.7)
www.reddit.com/user/shawnyv

183 cm, 80.7 kg (BMI 24)
www.reddit.com/user/alittamnayr (Ryan Mattila)

All three men have impressive muscles, and none of them look 'too skinny'. So you can see that the mainly male concern that a normal BMI might make you look weedy is totally unfounded.

Being overweight doesn't impede me in anything

I can't speak for millions of overweight people, so this chapter is exclusively about my own fat logic.

For a long time, I convinced myself that being overweight didn't impact particularly negatively on my life. I'd suppress the panting as I climbed the stairs, so I could tell myself that I had no problem walking up three floors. Once at the top, I would sometimes pretend to cough or laugh to hide the fact that I was out of breath.

I recently watched a video on YouTube in which an obese woman climbed 100 steps, and I felt like I'd been transported back to my earlier life. The woman's gait became more laboured from the 20th step, then she started to laugh, until her wheezing eventually became too obvious to hide, and she joked about how hard it was. The last 40 steps were pure torture, but at the end of her ordeal she was so proud of having shown how physically fit she was.

I realised I didn't know subjectively whether 100 steps were a lot or a little to climb, so I decided to do an experiment and find out for myself how difficult it was to walk up 100 steps for me, with a body weight of 67 kg and after six months of regular physical exercise (though it was also just a week after I'd undergone abdominal surgery). The result: it didn't even make me out of breath. So I kept climbing. I gave up from boredom after 600 steps.

Our society makes it very easy for us, unfortunately. People who exercise regularly are seen as 'fitness freaks', 'sports fanatics', or similar, while 'normal' people are the ones who lead physically inactive lives. For adults, at least, it's seen as abnormal to do a lot of sport. Most people 'don't have the time', or they admit openly that they just can't be bothered. And that's absolutely nothing to be ashamed of — at least, that used to be my opinion. When I was still very fat, many normal-sized or less fat people (I didn't actually know anyone who was fatter than me, so they were all less fat, to be honest) supported my sedentary lifestyle by pointing out how unfit they were, too. When I started exercising, most of them made comments like, 'I couldn't do that.'

Two years ago, the Mayo Clinic published a study (Archer et al., 2013b) that caused quite a furore because it showed that the average obese person does very little physical exercise. Obese women were found to do an average of only one hour's intensive exercise per year; obese men did under four hours a year. What media reports about the study usually failed to point out, however, was that women of normal weight were found, on average, to do only around 11 hours of intensive exercise per year (although 'intensive' exercise included only very strenuous activities that caused a considerable increase in heartrate, so, for example, a gentle bike ride through the countryside wasn't counted).

Considering that even 'normal people' are pretty unsporty, there will always be someone to compare yourself to so as to persuade yourself that you aren't actually that unfit.

Now that I can compare the abilities of my well-trained body (and I'm absolutely not athletic or super-fit) to my abilities before, I've come to realise how far below optimum my fitness level really was, and how ordinary the things I used to be proud of are. For example, I considered myself to be relatively strong. I was proud of my ability to carry 30 kg rocks, and I thought it was definitely thanks to my 'stout build'. Now, after weight training, I can carry the 45 kg kerbstones at our home (which I didn't even used to be able to pick up) with ease, even though I'm now a lot less 'stoutly built'. The fact that fatty tissue doesn't give you any strength, and, on the contrary, actually hinders movement, took a while to sink in. It would never have occurred to me that a fit woman weighing 50 kg could be stronger than me, weighing in at 150 kg. For me, a large body mass meant strength. In view of the fact that, according to Pellegrinelli et al. (2015), fatty tissue can even atrophy muscle mass, this was, of course, doubly wrong.

Although I had fewer illusions when it came to my stamina and general fitness, I was still very wrong about myself. One day, when my car refused to start after work, I decided to walk. Arriving home after the 45-minute walk, my feet were hurting, and I was so exhausted that I had to lie down on the couch and rest before I could do anything else. The sad thing is that I didn't even see it as a warning sign. In fact, I was proud of my achievement. I thought it wasn't half bad that I'd managed to cover that distance 'easily'.

The other day, on a Sunday when most shops are closed, I really wanted something sweet, so I spontaneously walked the 12-kilometre round trip to the nearest petrol station and bought some chocolate. For me, this is pushing my limits in a pleasant way. A real athlete might laugh at my 19 per cent body fat, but on normal charts I'm somewhere between 'athletic' and, more usually, 'below average' for women of my age group.

I expect most readers will fall somewhere between being fit and weighing 150 kg, and so it could be tempting for you to say, 'Everything might be awful when you weigh one hundred fifty kilos, but I'm a long way from weighing that much.' Of course, that's true. A BMI of 50 is extreme and can't be compared to a slightly obese BMI of, say, 32. But even if many of the problems that come with weighing 150 kg are not as bad for someone who weighs 100 kg, there's no doubt that the beginnings of those problems will definitely be there.

In retrospect, with the comparison between before and after now available to me, I can see there were many symptoms that I didn't recognise at the time, or that I considered to be normal. My many years of obesity had robbed me of the ability to imagine what a normal body might be capable of, and what it would feel like to live in a healthy body.

One change that surprised me has to do with sleep. Ever since I'd first met my husband, I'd always needed one to two more hours of sleep than him. I slept about nine hours a night and saw myself as a late riser and definitely not a morning person — I was someone who needed time to get going in the morning. I now sleep for about seven hours a night and am fresh and bright in the mornings.

I've turned into one of the people I used to find so strange, who got up at 6.00 a.m. of their own free will. I'd thought of it as something to do with personality or character, and I never made the connection with body weight. I assume that my lower sleep requirement is thanks to the fact that my sleep is now more restful and restorative because I no longer snore and can breathe more easily.

Another change is that nowadays I truly enjoy physical exercise. Whenever someone used to say that they found active sports fun, I would dismiss it as twaddle because it was such an inconceivable idea for me. As it happens, one reason why it's easier to be thin than to get thin is that a fat body is significantly less energy efficient. Although it burns more calories at rest and during exercise than a thin body, it's much more difficult as a fat person to do the same amount of exercise as someone who's slim. When I weighed 150 kg, a ten-minute warm-up on the exercise bike was like torture to me. And I was frustrated by the fact that it only burned 60 kcal. I now pedal faster and with more power, and I can burn about 120 kcal in the same time. An hour on the cross-trainer feels like relaxed movement and, on the setting I prefer, burns 600 kcal.

As I've already stressed, I'm talking about myself here. I'm not saying that there aren't any overweight people who do a lot of exercise. But I can imagine that lots of people fall prey to a similar kind of distorted thinking as I did when I used to consider even relatively normal things to be great sporting achievements. A short walk with a weight of 150 kg was harder than an hour of endurance training with a body weight of 65 kg. But although I'd pant and sweat more, the number of calories I'd consumed was much smaller.

The same applies to the achievement that one of the US leaders of the fat acceptance movement, Ragen Chastain, claims makes her an 'elite athlete' — with a morbidly obese BMI. In 2013, she ran a marathon and published an article about it with the title 'My Big, Fat, Finished Marathon'. The essence of the article was that after five months of training, she covered just over 40 km in 12 hours and 20 minutes. Now, I'm not saying that it's not an achievement for a severely obese person to walk the entire length of a marathon in one go. But her average speed of less than 3.5 km per hour is much slower than normal walking speed. The marathon had officially ended hours before she finished it, the stands all removed, the organising team gone, leaving just a few poor souls who had to wait for the final participant to finish before they could finally go home. The last participant to complete the race, several hours before Ragen crossed the finishing line, was a woman who was over 70 years of age. Here are more figures for comparison: the women's record for running the marathon is two hours and 15 minutes, and the average time for women is about five hours. In 2011, a one-hundred-year-old man completed the marathon in eight hours and 25 minutes.

More disturbing than Chastain's false evaluation of her own athleticism, in my opinion, is the fact that she complains in her article that one of the volunteer workers harassed her and treated her in a discriminatory way. All the volunteer had done, once she realised after several hours that Chastain would not be able to finish the marathon within the appointed time, was to ask her to abandon the race — for one thing, because she was simply worried about the health of this overweight woman, and for another, because some of the volunteers were being forced to stay much longer than planned because of Chastain. Ragen refused to quit, however, as the marathon didn't have an official closing time. She may have been within her rights to refuse, but then to be up in arms because she felt she was treated badly, when she had forced an entire team of volunteers to hang around for four extra hours until she had hauled herself over the finishing line, is a bit rich.

I think Chastain's example shows very clearly how far people's grip on reality can slip when they are severely obese for a very long time, and they are in an environment in which it's seen as *normal* to be severely overweight.

Please don't take my statements here as a lack of appreciation for sporting endeavours. But of course, everyone has to start from their own individual fitness level. When I weighed 150 kg and was more or less unable to move for six months, average sporting achievements were as likely for me as breaking Olympic records. In the first few months, I was proud of reaching various milestones, like walking for half an hour without stopping, or spending 20 minutes on a bike for the first time in years. They're things that an average person would never see as milestones, but that for me, at the time, were great strides. I think it's good to be proud of your own development and individual progress, even when it might not objectively seem that impressive. After all, for someone who's never smoked, going a day without lighting up a single cigarette is not much of an achievement, but for a habitual chain-smoker, it would be a great personal success. Doing the weekly shopping at the supermarket is normally no great shakes, but for someone who'd been stuck at home with agoraphobia for years, it would be a massive achievement

Pride in your personal achievements and improvements, however small they may be at first, is a great motivator, so it's very important. But declaring your own, below-average performance to be an objective record, and therefore to claim that any improvement is unnecessary, will only stop you — and others — from tackling the problem of excess weight and thereby reducing the risks to your health.

If I were thin, all my problems would be solved

While some people believe that their weight doesn't impede them in the slightest, some go to the other extreme and blame their weight *exclusively* for *every* problem in their lives. No partner? Of course, it's because I'm fat so no one wants me. Unhappy at work? Of course, it's because no one sees past the weight and appreciates me for the work I do. If I were thin, my boss would see me differently. No fulfilling hobbies? Of course not, I could never join a yoga class with this fat butt — everyone would laugh at me.

I made this mistake in my twenties, when I starved myself down to a normal weight within a few months, and then realised that I was still the same person, only thinner. Neither my life nor my personality suddenly became better. I guess that was one of the reasons I thought being fat was a rational decision and being thin just wasn't worth it. I didn't realise that a healthy body is just a means to make the most out of your life, but it doesn't bring automatic fulfilment. Being thin isn't the meaning of life. It doesn't make you happy. It's a little like money in that way. If you're poor and don't have money, your lifestyle is limited, and there are a lot of things you can't do. But just having money in the bank doesn't make you happy if you don't enjoy it by using it to do something that benefits you in some way.

On the one hand, losing weight per se hasn't changed much for me, but on the other, the consequences of that weight loss have led to many changes. The fact that I now take pleasure in sport and exercise has opened up an entire spectrum of new interests to me that would have been out of the question before. A few months ago, my husband and I went on a cycling holiday by Lake Constance. I've discovered climbing as a new hobby — and pilates, too. My physiotherapist has become my gym buddy, and we now meet once a week for coffee and weight training. The amazing thing about all this is that these activities are now fun for me. Through this, I've come to realise that I have gained so much, and that being thin doesn't mean a life of constant deprivation, as I used to believe because of my previous experiences.

So, what was the difference between my disappointing date with thinness in my twenties and my new relationship with being thin now? It probably comes down to expectations.

Before I lost weight in my twenties, my expectations were, firstly, that my life would change completely when I was thin, and secondly, that I would be able to revert to my previous eating behaviour because as a thin person, I would be able to 'eat whatever I wanted'.

Because I starved myself down to a normal weight without a real plan in mind, I also had no concept of how my long-term eating and exercising behaviour should look subsequently. So for a while, I swung between overeating and fasting, with my weight fluctuating between 68 and 73 kg. But I never settled into a stable pattern of eating behaviour. In addition, things like stress at university and dissatisfaction with my social life didn't start to improve automatically. In fact, concentrating so hard on my weight made it difficult for me to change anything in other areas of my life.

It then felt liberating to change my priorities and concentrate on my studies and my doctorate, start a relationship, and build a life, and I dismissed the issue of weight, even though that meant that it once again got out of hand. My previous experience had reinforced the idea in my mind that maintaining normal weight would be a constant battle for me and would leave no space for anything else in my life.

Instead, my experience now is that being physically fit and healthy *gives* me space, opens up more opportunities, and makes my day-to-day life easier, because I have more energy and am less limited in what I can do. One crucial factor is that my knowledge of bodily processes, nutrition, and fitness means that I'm able to plan my nutrition more sensibly. It no longer includes panicky counter-regulation measures but is more uniform, and it doesn't occupy as much space in my life. I still count calories and weigh out my food, but I'm also able to leave that routine at home when we go out to eat, or when we're invited to friends' dinner parties. Food is no longer associated in my mind with worry or fear, and, for my part, I find it helpful to have clear guiding principles.

Another difference is probably exercise. Throughout my life, I've made several attempts to engage with sport, including the period in my twenties when I was of normal weight, but I never managed to throw off the idea that sport was 'just not for me'. I never got to the point where sportiness became part of my self-image. I always saw it as something alien, which I could do if I forced myself with a great deal of discipline, but which could never be enjoyable. For a while, I thought jogging and cycling were 'okay', but I never had the experience people talked about of really wanting to exercise.

Weight training was a great help for me in this. Training with weights produces rapid results in the form of muscles, while endurance training hardly makes any visible difference (although it improves your fitness, it's not as visibly obvious). It's much easier to think of yourself as athletic when you can see your muscles. I guess this difference is purely psychological, but for me, training with weights was a great help. Another contributing factor is that when all the muscles in your body are strong, new sports are much easier right from the start, and that, in turn, allows you to develop an 'athletic self-image' more quickly.

[Female fat logic] Weight training? No, thanks. I don't want to be muscly

In our society, weight training for women is seen as something negative. People immediately picture massive female body builders, with huge bulging muscles and broad shoulders, who 'look like men'. I guess everyone's seen these kinds of extreme female body builders in the media, and it seems most people consider them to be unattractive.

I think one of the reasons for this is that our ideal of beauty is based on a 'healthy appearance' and therefore both being very underweight and being very overweight (irrespective of whether it's due to fat or muscle mass) is perceived as unattractive by most people. It is a fact that women naturally have less muscle mass than men. The blog Alles Evolution gives the following breakdown of the average physical differences between men and women:

- Men weigh around 15 per cent more than women;
- Men are 15 cm taller than women;
- Men's upper bodies are 40 to 50 per cent stronger;
- Men's lower bodies are 30 per cent stronger;
- Men have a 30 per cent greater lung capacity relative to their body size;
- Men's elbows and knees are around 42 to 60 per cent stronger;

- Men's skin is thicker and greasier;
- Men have more blood corpuscles;
- Men have more haemoglobin and so can store more oxygen;
- Men's hearts are 10 per cent larger relative to their body size;
- Men have stronger bones;
- Men can dissipate more heat, because they have more sweat glands;
- Men's wounds heal more quickly.

Women, by contrast, naturally have more body fat and more elastic tissue, as their bodies are designed to provide for a baby in the womb. So, unless she intervenes massively with male hormones, a woman will never be able to build up anything near the amount of muscle mass a man is able to gain. It's biologically impossible. Of course, when it comes to losing weight, that's a real disadvantage because muscle mass is the only significant way of influencing the body's resting energy requirement.

Men already have more muscle mass in their 'raw state', but they also build up muscle mass more quickly with weight training, giving them a great way to speed up their weight loss.

Women, on the other hand, who are already at a disadvantage, are additionally influenced by societal nonsense advising them to avoid lifting any heavy weights because they might accidentally heft a big dumbbell and wake up the next morning with shoulders like Arnold Schwarzenegger.

The truth is, a 'defined' figure, with muscles visible beneath your skin but not especially pumped up, is about the maximum women can achieve, even with intensive weight training — unless they resort to other ways of gaining muscle mass. The 'danger' of suddenly becoming 'too muscly' is so slight as to be negligible. Especially when you think about how much effort serious female body builders have to invest in their bodies.

But women are still encouraged to take up more 'feminine' sports like those based on endurance rather than strength, or like aerobics without weights. There's nothing wrong with those sports, but they do little or nothing to stop the body looking 'flabby' (unless the woman has very little fatty tissue and the smaller muscles are visible) or to address the fact that even quite light women have a relatively large amount of body fat and not much muscle mass.

Another little-known fact is that that common female bugbear, cellulite, isn't just a tissue problem, but also a muscle problem. When muscles atrophy, the tissue on top of the muscle becomes detached and dimples in an unattractive way. Cellulite isn't an unavoidable fate for women as they get older. It's a result of flabby muscles and too much fatty tissue, and it's reversible — at least to some extent.

What's sad is that this particular ideal is held up as desirable — a taut, defined, slim body — but at the same time, women are discouraged from doing what's necessary to attain that ideal, i.e., building muscle mass. Instead, they're encouraged to work senselessly for it and spend hours running on a treadmill or cycling on an exercise bike with low resistance (so that their thighs don't get too strong). I see a lot of these kinds of women at the gym, who then move on to the leg press where they 'work' with 30-kilo weights. Pressing such a low weight is so unchallenging that getting up from the leg press afterwards actually provides more training. It's why a lot of women start to get frustrated by the fact that they see no difference in their bodies, even though they're going to the gym several times a week.

The worst-case scenario with this kind of work-out is that the woman will then fall prey to the mistake described elsewhere in this book: she will overestimate the number of calories she's burned during a training session, will eat accordingly, and as a result her fat will increase rather than decreasing —meanwhile, she will mistakenly conclude that after the first month of training those extra two kilos stem from 'building up muscles'.

Weight training is advisable not just because it's good for your appearance and a good way to lose weight, but also because it improves your general feelings about your body. To judge from my own experience at least, it boosts self-confidence because it makes day-to-day tasks easier to perform without help. While we were doing up our old house, I was always frustrated at having to wait for hours for my husband to turn up because I wasn't strong enough to do some jobs myself. Now I'm able to do a lot of those things by myself, and that independence is liberating.

The improvement weight training makes to body posture also helps to boost your self-confidence. The muscle mass built up by weight training can help you to have a more upright posture, and that, in turn, has a positive influence on your mood and general attitude. According to Cuddy (2012), body posture has a big influence on self-confidence. For example, people are far more successful in job interviews when they spend a few minutes beforehand consciously standing, or sitting up, as straight and as confidently as possible.

And, yes, weight training is also very good for your health. If you're beginning to think I sound like a fanatical supporter of weight training, you might be right. I do think weight training is extremely useful — for both genders, of course. I've addressed this chapter to women specifically, because I've never heard a man express a fear of becoming 'too muscly'.

I had a memorable experience of this societal nonsense with someone who really should have known better: the physiotherapist who treated me after my knee operation. He strongly recommended to me that I should have a second knee operation because apparently I would never be able to achieve the build-up of muscle that was recommended by the orthopaedic surgeon in order to stabilise the joint. The physiotherapist's words were something like, 'That would involve having to do weight training at least twice a week — and wouldn't you rather look good?'

If patients are discouraged in this way, rather than being *en*couraged to at least give it a try, it's hardly surprising that the success rate is correspondingly low. It's like me telling one of my psychotherapy clients who's afraid to leave the house because of his agoraphobia to find some good food-delivery services rather than work on his fears.

After my prescribed six sessions with that physio-therapist were up, I took my next prescription to a difference practice where there was a strong focus on muscle-building exercises. The physiotherapist there was the one I became friends with and whom I now meet up with regularly so that we can train together as friends.

That experience brought home to me again the extent to which lack of activity has become the norm in our society. Half an hour's weight training three times a week should be something that is generally encouraged — and not just in order to avoid a second operation.

This demotivating attitude to physical exercise is probably not purely a women's problem. Overweight people in general are often told that there's no chance of them ever getting properly fit. The only difference is that at least men have an ideal to strive for, whereas women are told that it's not only impossible to achieve, but also undesirable.

I've gained weight since I started exercising, so it must be muscle mass

For a recent study, 81 women walked on a treadmill for half an hour three times a week for twelve weeks. Measurements of their body fat showed that 55 of the women gained weight during that period — and the gain was in fat, not muscle mass (Sawyer et al., 2014). This is a common pattern because we tend to seriously overestimate the number of calories we burn by exercising, and sporting activity stimulates your appetite.

The results are corroborated by studies carried out by Finlayson et al. (2011) and King & Blundell (1995), in which participants in exercise programs reported having a highly increased appetite for fatty foods right after exercising. It seems people fall into different types when it comes to exercise: some are less hungry after an exercise session, while about half the population find their appetite enormously stimulated after exercising, leading them to replace the energy they have burned (or to consume even more).

Physical trainers say it's possible for an average man to build up as much as 23 kg of muscle mass over his lifetime using natural methods(!), with many years of continuous weight training. For women, it's possible to build up around 11 kg of muscle mass. As it happens, that also means it's basically impossible for a woman with a low proportion of body fat to be in the overweight BMI range because of naturally gained muscle mass.

The speed at which muscles grow is also surprisingly slow: according to the fitness expert Lyle McDonald, men can build up around 1 kg of muscles per month in the first year, under optimum conditions. In the second year, they can build around 500 g per month of muscle, in the third year, around 250 g per month, and in the fourth only 2 to 3 kg for the whole year. For women, the figures are about half those of men.

The nutritionist and sports educator Alan Aragon gives a guiding estimate for beginner male weight trainers, telling them to hope to gain around 1 to 1.5 per cent of body weight in muscle per month. Once again, the figures are about half that for women.

This means that if a man puts on more than 1 kg during a month of intensive weight training, and a woman puts on more than half a kilo, the gain is probably due to fat, and it's time for them to take a closer look at their calorie intake.

But there's no need to panic right away. Muscles can sustain many tiny injuries at the beginning of an exercise program, and that can cause the tissue to swell up and therefore weigh more. So if you suddenly find that you weigh 1 to 2 kg more the day after starting to exercise, there's no reason to worry or to give up sport immediately: that additional weight will disappear again, and it's nothing more than a sign that your muscles have been put under strain.

What's more, the pattern of putting on fat after exercise described here will only kick in if you eat intuitively. If you are alert to the appetite-stimulating effect of exercising, you can avoid falling into this typical trap. Just because you have an appetite, it doesn't mean you have to eat something. And if you know you're likely to get hungry after exercising, you can plan your day so that you can eat a meal after your sports session.

Before I try to lose weight, I need to find out why I'm fat

I can imagine readers will be surprised that I, a psycho-therapist of all people, should declare root-cause analysis to be 'fat logic'. Of course, I'm not trying to say that there are no psychological reasons for unhealthy body weight. Ultimately, most people realise that being under- or overweight is bad for them. The same is true of lots of things: vegging out in front of TV when you should be tackling your tax returns, or ignoring the strange noises your car is making because you can't be bothered to take it to the garage this week.

A lot of people have the wrong idea about what psychotherapy is capable of. It's not like in the movies, where ten therapy sessions lead to a realisation that the client was sexually abused as a child and has since been overeating to give herself a 'protective barrier' of fat, and now that the reason is known, the fat shield falls away of its own accord. I'm not saying that those kinds of cases don't exist, but when 60 per cent of the population are overweight and one in five is obese (Mensink et al., 2013), there is reason to believe that not every spare tyre is the result of a tragic past or deep psychological problems. I am the last person to deter anyone from seeking help through psychotherapy if they feel the causes of their overweight condition are psychological in nature, but that kind of search for *one single* cause is often what stands in the way of change.

In psychotherapy it's also sensible to address an obvious problem right at the start, since insights often only come after change is made. If a client's behaviour serves to conceal another problem, it's often brought to light by them changing their behaviour patterns and seeing what comes up.

In an article by the British journalist and author Johann Hari (2014), he discusses cocaine, which is considered to be one of the most addictive drugs. In the 1980s, experiments showed that nine out of ten laboratory rats became addicted to cocaine when they were given water laced with the drug. But one psychology professor questioned those results, pointing out that the rats were kept in their bleak cages alone, with nothing to do but take the cocaine. So he built a little play park for rats, with toys, tunnels, good food, and lots of other rats to play with. He offered the rodents cocaine in that environment. He found that the happy rats were not particularly fond of the cocaine water and consumed less than a quarter of the amount taken by the unhappy rats. All the lonely rats became intensive drug users, while the happy rats did not, and did not die of cocaine overdoses.

In a follow-up experiment after 57 days, the scientist placed the severely addicted, unhappy rats into the nice, play-park cage. The animals displayed some withdrawal symptoms but stopped taking the cocaine water and were able to return to a normal life.

I find this experiment hugely fascinating and I think the results can be applied to other behaviours. Addiction covers a broad spectrum and doesn't always have to involve a physical dependency — just think of shopaholics or gambling addicts. Excessive eating doesn't always share the full set of characteristics of an addiction, but over a longer period, it does tend in that direction.

Eating can be a strategy for coping with feelings like stress, boredom, or loneliness. In a relationship, it can also be an expression of love and affection, for example, in the form of cooking for a partner. This only becomes a problem when eating is the *only* strategy for coping with those emotions, and a dependency on it develops in the absence of any other way of handling those situations — like the rats in their lonely cage, who had no alternative.

Of course, that doesn't mean every overweight person leads a bleak and dismal life. It just means that they may have no alternative for coping with certain situations.

I once had a client who was unable to kick his heavy smoking habit. His job was so stressful that his occasional cigarette breaks were the only opportunity to take a couple of minutes off, as his highly perfectionistic nature prevented him from taking any breaks *voluntarily*. Without the excuse of addiction, he would have continued working without a break until he dropped. Our therapeutic approach was to find strategies that would enable him to allow himself to have other types of breaks and to question the ingrained idea that he must always work to the point of exhaustion.

Behavioural patterns can also become vicious circles over time. Being overweight can become a barrier to finding other ways to gain satisfaction or relieve stress. When moving becomes difficult so that exercise isn't enjoyable, a lot of possibilities are out of reach. You don't need to weigh 150 kg and be unable to move for this to happen. Even an extra 10 kg can make physical exercise more difficult and strenuous, and less enjoyable, so that the overall experience becomes negative rather than positive.

This is compounded by the existence of several bodily mechanisms that make it difficult for us to change our habits at first: gut bacteria, which foster our preferences for certain familiar types of nutrition; the reward centre in our brain, which prefers the high-calorie food it knows; and our blood-sugar regulation system, which makes us want to take counter-regulation measures when our blood sugar level drops.

It's a mistake, though, to react to those initial difficulties by thinking things will never change and just giving up. Reprogramming both your body and your habits takes a few weeks or months, but once that's happened, the new lifestyle no longer feels like a permanent struggle.

The point is that removing the deeper cause is almost never successful. Someone who uses chocolate as a strategy to cheer themselves up, to relax, or as a reward isn't going to be able to give up chocolate by banishing stress from their lives forever. But it is possible to find alternative strategies that work just as well.

Some of those strategies have to be built up with hard work, until a point is reached where the pleasure of playing a certain kind of sport, for example, is comparable to that of eating a piece of chocolate. Initially, exercising may even *cause* stress, and doing it feels as if it warrants the reward of a bar of chocolate.

That's a normal reaction, though. Changing habits is hard work, and the phase in which it takes place is a difficult one. On the other hand, it's also a phase that can provide a great sense of achievement at the successes that the new pattern of behaviour brings. That's one of the reasons why I believe it's so important to gather as much basic knowledge as possible about nutrition, calories, and weight training, to stave off the danger of not recognising the positive results of that laborious process of change and therefore giving up, twice as frustrated as before.

Ideally, that difficult phase of adopting new habits will also include a lot of positive experiences. That may include watching the numbers on the scales go down, feeling the changes in your body, receiving positive and encouraging comments, or discovering fresh interests. It can provide the motivation to keep going. Without positive reinforcement, it would require almost superhuman self-discipline not to give up.

The successes must be in relation to the effort. The results of a slow, gradual change in lifestyle will come less quickly, but that kind of change is usually seen as less difficult. A radical change, on the other hand, needs to bring with it rapid positive results.

It can have a powerfully disheartening effect, when, for example, a very overweight man makes the decision to focus his effort on exercise, overcomes all his misgivings, and starts going to the gym five times a week in the belief that exercise is the answer. He might feel uncomfortable there and feel as though everyone is staring at him, and he might find the physical exertion very difficult. And then despite the massive effort he's putting in, his lack of general fitness means the results are rather modest. Although he gives his all for a whole hour, sweating and pushing himself, he only manages to burn off a measly 400 kcal. Doing what feels to him like an extreme exercise program five times a week, he manages to work off 2000 kcals, which is equivalent to about 300 g of body fat, which in turn works out at a little more than 1 kg in a month.

Losing just 1 kg for all that hard effort will probably lead him to feel very frustrated. It's also compounded by the fact that most people find endurance training increases their appetite, either making it even more difficult to eat less, or tempting them to replace the burned energy straight away — with well-meaning protein shakes, or in the form of a 'well-earned' reward. Because of this, the man may even gain weight rather than losing it, despite all the effort he's put in. For a month or so, he might be persuaded by well-meaning remarks telling him that he must be gaining muscle mass. But when success continues to elude him, he'll eventually give up, thinking, 'I invested so much and achieved nothing. Losing weight isn't an option for me.'

Counting calories is similar. Some people launch a half-hearted attempt to start counting, investing time and energy in it, but then too often slip into guessing (*I can't weigh that right now, but it must be about fifty grams … or maybe more like forty*), cheating (*It was just a little bit, I don't have to count it*), or forgetting (*I'll write down that bar of chocolate in my food diary when I get home this evening*). Eventually, the calorie-counting process becomes so inaccurate as to be useless.

In these cases, the only thing that helps is to make sure that the results match the investment. As I see it, the best way to achieve this is to concentrate 90 per cent of your efforts on your diet, and to really take it seriously. This is true irrespective of the size of your target caloric deficit. It doesn't matter whether you want to lose a lot of weight quickly and run a big deficit, or you're aiming to shed 2 kg or less per month and only need to cut back a couple of hundred calories — both strategies are justified. The important thing is not to sabotage your own efforts by falling into the abovementioned traps and decimating your deficit every day.

People who are already physically fit can earn their deficit with exercise, seeing as that investment is considerably less challenging for them and yields greater results. In some cases, they won't even need to use dietary methods to save on calories. Whether you find it better to do more exercise and go hungry less, or do less exercise and go hungry more is simply a matter of individual preference.

In my opinion, people who aren't physically fit should not initially think of exercise as a way of losing weight, but should start by concentrating on getting their body into a state in which it is able to do sport at all. As an initially unfit person, any activity to build up muscle mass or improve fitness should not be seen in terms of burning calories (and should not be included in a calorie-deficit calculation), but should be seen as an investment in future fitness.

I'm doing everything right, but I've stopped losing weight

During the weight-loss process, a lot of people reach a point where the needle on the dial of their scales gets stuck on one number and stubbornly refuses to go below it. It's often the point at which well-meaning people around them start talking about starvation-mode metabolism or asking whether they've had their thyroid function tested.

The first question in these cases is a trivial one: how big is your daily deficit and *is it up to date*? A calorie-counting app will calculate a new daily energy requirement every time you enter a new body weight. Those calculating by hand must, of course, remember to recalculate their requirement regularly. If I weighed 10 kg more, for example, my base requirement would be 100 kcal higher. In the same way, the body's base requirement grows less over time as you lose more weight. When your daily deficit is small, you can, of course, still one day reach a point at which your energy input and consumption are in balance and no more fat is being broken down — which would happen more obviously with a large daily deficit. But that point approaches slowly and gradually.

Medical conditions like hypothyroidism can only make themselves noticeable as a complete standstill in weight-loss when your calorie deficit is small. With a daily deficit of 1000 kcal, no one is going to suddenly stop losing weight because of an underactive thyroid, which only slows the metabolism by about the equivalent of 100 to 200 kcal. At most, people with hypothyroidism might notice that they've lost only 3 kg rather than the expected 4 kg after several weeks.

Not using up any fat reserves despite doing everything right (i.e., eating considerably fewer calories) is an impossibility. A body cannot conjure up energy out of nothing, and if less energy is put in than it requires to function, it has to fall back on its reserves. Thus 'inexplicable' plateaus always turn out to have an explanation.

Once fluid retention has been excluded as an explanation, and it becomes clear that less fat has been burned than should have been, the explanation may turn out to be one of the following:

1. Overestimation of the base calorie requirement. Formulae for calculating the base requirement can offer a very good initial estimate, but they shouldn't be relied on to be 100 per cent accurate. If, for example, your body has less muscle mass than average, its base requirement could be considerably lower than the figure produced by the formula. Even with an inaccuracy of only 5 per cent, the difference can be as much as 100 kcal per day, which works out at half a kilo per month. So, although those calculators are useful at first, as a guide, it makes more sense later to estimate your actual requirement on the basis of your personal data (deficit and actual weight loss).

2. Counting exercise twice. If you enter a certain number of hours of exercise per week into a calculator, those hours will be credited to your average energy requirement. If you then either do less exercise than you predicted, or you enter those same hours of exercise again by hand, the amount of energy the calculator will show you consumed will be far higher than the actual value.

3. Overestimation of exercise. Exercise machines and calorie calculators show the total number of calories burned during the entered time. But we would still have burned some calories if we had stayed on the couch rather than running on the treadmill. If I enter into a calculator that I spent 24 hours lounging on the couch watching TV, it will show that I burned about 1900 kcal — my daily resting energy requirement. But if I enter *that* figure into my app, it will calculate a daily requirement of 3800 kcal for me, which is total nonsense, of course. It's important to remember always to subtract the number of calories that would have been burned in any case. For me, with a daily requirement of 1900 kcal, that would be about 80 kcal an hour. So, when my exercise bike tells me that the number of calories I used in an hour's training is 500 kcal, I can only count at most 420 kcal of that as extra calories burned.

4. Inaccurate calorie information. Studies, including, for example, Urban et al. (2010), have shown that the energy information given on convenience food packaging is around 8 per cent lower than the

actual calories it contains. For restaurant meals, that discrepancy is as much as 18 per cent. That means that people who eat out a lot or who rely a lot on pre-packaged food need to be particularly careful and bear in mind that they should err on the side of caution when counting their calories. A 230 g package of pre-prepared food should really be counted as 250 g and when you enter the calories into the calculator, you should enter the higher value.

5. Inaccurate weight information. For a long time, I trusted the weights given on food packets and accepted them as correct ... until I noticed that some of them couldn't be right — for example, when a 125-gram pack of mozzarella only turned out to produce two daily portions of 50 g each. Weighing foods myself revealed that the information on the packet often deviates greatly from reality in one direction or the other. So it's a good idea to weigh your food out yourself, especially in the case of high-calorie foods, rather than relying on the information provided by manufacturers.

6. Forgotten calories and ignored calories: it wasn't until I'd been losing weight for six months that it occurred to me to check whether the vitamin supplements I was taking contained any calories. And so they did: about 15 to 20 kcal per fizzy tablet. Multiple overlooked sources of additional calories can accumulate and affect your deficit. It's also important to factor in tiny bites of food that might seem too ridiculous to record. A sample of cheese at

the supermarket, a sip from a friend's glass of juice, a single nut ... Those who can't be bothered recording such tiny amounts should instead leave enough space for them in their calorie deficit and, if necessary, increase their target deficit accordingly.

If there is no progress on the scales for several weeks despite a calorie deficit, it's a good idea to run through the points listed above once again to identify any possible sources of error. Once all these points have been excluded, then it makes sense to take your search for the reason to your doctor.

There are also theories that claim that fat cells sometimes retain additional fluid during weight loss, making parts of the body feel 'squishy'. I haven't been able to find any official evidence of these theories — only online texts on some fitness and weight-loss websites. They describe the effect as being similar to normal fluid retention, including that the fluid is eventually 'released' and suddenly disappears within a few days. This is then noticed as a large and sudden change in weight (the 'whoosh effect'). The claim is that the whoosh effect can be triggered by, among other things, a very high-carb meal or by alcohol.

I was prompted to look into this theory a little because I myself went through phases when my weight loss stagnated for several days despite my large caloric deficit. I wasn't able to recognise this whoosh effect in me as being a result of carbohydrates or alcohol. But, now that I have maintained my weight at around 65 kg for several months,

I have noticed how powerfully my weight is affected by my hormones. At a certain point in my menstrual cycle I usually lose around 4 kg over a very few days, then regain 2 kg slowly over the next two weeks, and then put on the other 2 kg again quickly within a few days. This is completely independent of my caloric balance, which remains relatively constant.

On the basis of my experience, I wouldn't consider an absence of weight loss to be a real stagnation unless it lasted longer than two weeks.

And even then, that kind of stagnation is not 'inexplicable', but must be explained by one of the following two reasons:

1. you have a caloric deficit but your fluid retention is preventing it from being seen on the scales;
2. your calorie intake is too high.

Counting calories is stupid — you can never count them accurately, anyway

Counting calories is stupid. The calories contained in fruit depend on how ripe it is, and the information on food packaging is often out by several per cent. It's completely impossible to determine the exact number of calories in what you're eating. Counting calories is a total waste of time.

Interesting ...

... I'd like to know more about that, but I have to go now, and who knows when we'll see each other again ...

What? We're seeing each other tomorrow.

Tomorrow? Don't be stupid! I don't make plans.

Why would I? You never know whether or not you might get stuck in traffic, or your previous meeting might overrun. Scheduling is completely impossible.

I can't lose weight because I'm depressed

Depression is not an issue to be taken lightly. It's often difficult for people to even admit to themselves, let alone to others, that they're suffering from depression. It's exactly because of this that I feel it's important to offer some information about the link between depression and being overweight.

In truth, many kinds of mental illness affect weight. In the case of depression, changes in appetite and weight gain or weight loss are among the possible symptoms of a depressive episode. For those affected, this makes controlling their weight difficult, and large fluctuations in weight are not uncommon. Many people rightly say that the top priority during a depressive episode should not be to focus on weight, but rather to treat the depression.

One of my proofreaders agreed to tell a little of her own story about weight and depression. This is one of many examples of how depression and body weight can be related:

> I am 30 years old, female, and fat. The official term would probably be obese. Exactly when I got this fat is a difficult question to answer. I think I was this way in primary school. During my school years, my weight problem stayed within reasonable limits. I wasn't really fat, just a bit chubby. Especially

around my stomach. (In the final year of high school, people used to ask me if I was pregnant … so you can imagine how my stomach looked …)

When I finished high school, I was only 75 kg, with a height of 1.72 m — not really that overweight. If it hadn't been for my stomach, you wouldn't have noticed anything. Or it wouldn't have been that obvious, anyway. Then because of problems in my romantic life and bullying during my apprenticeship, the situation changed a lot. I now had enough money to buy whatever I wanted. Chocolate, crisps, coke. The more I was bullied and the more unhappy I was with the situation I was in, the more I ate. That's classic comfort eating. I shovelled all my frustration into myself, in the form of food. By the end of my apprenticeship, to put it in numbers, that was an extra 35 kg!

I was well aware of it. In part because almost every day since I first started school, people around me had rubbed my nose in it — telling me that I was too fat and that I ought to lose some weight, and reminding me that no one liked me because I was so fat, and that I'd never get a boyfriend: 'Who's ever going to want such a fat, ugly monster?'

That last section is, of course, important: the reactions of those around us can have a huge influence on our self-esteem, and low self-esteem is in turn one of the symptoms of depression. Years of bullying in particular can cause depression, or exacerbate it, and weight is an easy target. I had similar experiences as a fat child. I was unpopular in high school and was attacked for my weight. At the same time, the most popular girl in our class was as fat as I was. But she wasn't bullied. Being fat feels like you're walking around with a bullseye on your back for all to see. It's not a problem as long as you're accepted and popular and the people around you don't target you. In enemy territory, however, fatness is usually the first thing bullies will go for. But that doesn't mean being thin will necessarily protect you from their barbs.

> So, the stronger the pressure got, and the unhappier I was with the way the people around me were treating me, the more weight I put on. I didn't care any more. Of course, I'd always wanted to be thin, pretty, and popular. Who doesn't want that? But I'd never had the motivation to really wrestle with it. It was too much like hard work, and I was already fat anyway, so why bother paying attention to my weight?
>
> In my early twenties, I made some half-hearted attempts at losing weight. At one point I tried out a product that had been hyped in advertisements. Plant-based drops — can't remember exactly what they were. They cost me just over 100 euros for 50 ml. Of course, they made no difference. I'd have been better off throwing the money straight out the window — at least then I would have been able to pick it up again later from the ground …
>
> Another attempt was a four-week weight-loss

program at a gym, including nutrition counselling. Naive as I was, I not only let myself get talked into the program, I was also persuaded to sign a two-year contract. I'm sure you can imagine how it ended. After six months, in which I lost almost 10 kg and built up muscles I didn't even know I had, I stopped going to the gym. I didn't enjoy it. I had been coerced, and now I rebelled. I blocked out all comments about my weight or appearance and chose to focus on phrases like 'Fat people are cuddly' or 'Built for comfort, not for speed'.

And I still didn't care. I'd resigned myself to my protective armour. Although I absolutely hated it, and I still do.

This part of the story also shows the extent to which depression and obesity can be mutually dependent. In a sense, thinness becomes synonymous with social acceptance and popularity. At the same time, it also reinforces a depressive thought pattern: 'I'm not good enough. I must change and try harder to be accepted.'

A healthier response is to rebel against that pressure and to reject that kind of thinking.

On the other hand, massive external pressure has also led to a perception of losing weight and being slim not as something that is good for an individual's *own* wellbeing, but rather as a *capitulation* to social pressure or even *resignation* and *conformity*. Something you don't do for yourself, but for others. And if you succeed in the end, there can only be one of two possible outcomes:

1. Being thin doesn't change anything. Those around you are still critical, but they now criticise other things. Your 'only chance' of gaining acceptance is then gone. As long as you were fat, at least you still had the hope that losing weight would be a way to achieve happiness.

2. Being thin does actually bring with it social acceptance, friendships, or a relationship. But this also confirms the underlying assumption: 'I always have to make an effort to be liked. No one likes me just as I am. I have to fulfil certain conditions and keep working hard.'

Neither of these is ultimately a positive way out of depression. The problem here is that the physical condition is exclusively associated with its external social impact, and physical connections and consequences are hardly perceived at all, or fade into the background as secondary concerns.

> Soon after I turned 24, I was diagnosed for the first time with depression. I have been on permanent sick leave for four years because of it. For a little over a year now, my weight has remained within a range of 110 to 119 kilos. I can't climb stairs without arriving at the top puffing like a steam engine and drenched in sweat. Any physical exertion leaves me totally worn out. My back pain has become so constant that I only notice it when it gets particularly bad, or when I move in

the wrong way. My skin is blemished, and I also suffer from hair loss. Of course, I also have joint pain, aching muscles, shortness of breath after physical exertion, sensitivity to heat, moodiness, and presumably snoring and hypothyroidism. In a nutshell, I have now accumulated almost all the physical ailments mentioned in the book. This is all perfectly compatible with my depression. They each feed off the other. Obesity and depression. The classic spiral. But in my case, it spirals upwards with my weight. Getting increasingly worse.

This description illustrates how the physical consequences of being overweight are definitely experienced as burdensome, but that a lot of people perceive them as additional, rather than contributory factors to depression.

The woman who wrote this story then got to the heart of the matter of physical health:

For me, physical health would be a positive side-effect of losing weight. The same goes for the fact that it would help me get rid of my depression and be confident enough to tell people to their faces that I don't care about their opinion. At this point in time, that's about as possible for me as it is for a penguin to fly.

My reason for writing this chapter is to highlight the way many people seem to separate the physical and the mental, and assume that the two can be controlled independently of each other.

People with depression who are overweight tend to see their obesity as the cause of several of their other symptoms of depression: low self-esteem, (social) withdrawal, and a reduced ability to enjoy things. Often an aversion to doing certain things is associated with a fear of negative reactions due to being overweight. During a depressive episode, even positive activities like going out for an ice cream with friends are an emotional challenge, and that's made worse by the fear of negative reactions. An excessively self-critical attitude combined with actual 'funny looks' from strangers, or worse, mean that a tendency to dwell on imagined horror scenarios (*What if I'm sitting there with an ice cream and some teenagers walk by and they start jeering at me?*) can become a major problem.

It's important to point out here, though, that although these humiliating experiences play a part in some people's depression, they are by no means necessarily the factors at the root of the condition (though they may be perceived as such). They can change with weight loss, but that doesn't necessarily mean the underlying pattern for the depressed person will change.

If, for example, fear of humiliation or rejection is the reason why you can't bring yourself to tell someone to their face that they annoy you, then being overweight is only one possible opening for humiliation. The underlying fear of rejection is still there, and there will always be other possible points of attack. In these cases, the solution can't be to get rid of all possible targets for attack, but rather to deal with the fear of being attacked. So instead of saying to yourself, *I can't say what I think until I'm sure the other person won't be able to respond by insulting me*, it's better to think, *I can say what I think, and if they react badly, that's not my problem because I'm not governed by whether they think I'm thin, or beautiful, or a nice person.*

Some readers who are affected by depression might now be thinking resignedly, *Great! So my hopes for any improvement in my depression are pointless, and it actually doesn't matter whether I lose weight or not, because nothing is going to change, anyway.*

But that's not right either!

The fact is, body and mind are very intensely connected to each other and not at all separate. In the chapter on obesity-related illnesses, I've already covered the links between obesity and depression — like the fact that fatty tissue can contribute to depression by promoting inflammatory processes in the brain.

At the same time, there are direct connections between depression and dietary and physical behaviour. They can strongly influence each other.

The fact is that many factors that lead to obesity are also problems in the context of depression, and vice versa: things that help to combat obesity also help in the treatment of depression. In many cases, it's a mistake to treat depression as a mental-health problem in isolation, without looking at physical-health aspects.

For one, there's the direct influence that stems from the reduction of fat tissue, which reduces the inflammation markers in the blood, and which in turn can have a positive influence on brain chemistry and depressive symptoms.

What's more, a recent analysis by Josefsson et al. (2014) of several studies on endurance sports confirms what has been known for a long time: endurance sports have an immensely positive effect on depression, even exceeding the effects of 'traditional' treatment methods like antidepressants. For this reason, exercise is the first choice of treatment for mild to moderate depression, and a good complement to medication therapy for severe depression. So this means that part of a psychotherapist's job is often to work with patients to help them (re)discover sport, which can of course be difficult during depression because of reduced energy and drive. It is a way, though, to break the vicious circle of lack of motivation, inactivity, and the worsening of the symptoms of depression. Those who don't manage to start exercising straight away can achieve a lot even by simply walking: the symptoms of patients with depression were improved by taking 50-minute walks in nature (Berman et al., 2012).

Other studies, for example by Doyne et al. (1987), have shown that along with endurance sport, weight training has a similarly good effect in combating symptoms of depression. Again, it isn't necessary to go to a gym, as smaller exercise sessions at home are enough. The advantage of this is that smaller units of five minutes per day are a much lesser hurdle, even for those whose depression is strongly associated with motivation and energy problems.

Another important factor is nutrition: I've already discussed how a diet containing too many simple carbohydrates, like sugar and white flour, can cause big peaks in blood sugar levels, which then plummet again. These blood sugar, and hence energy, 'downs' can be felt as even more stressful than normal by patients with depression, so following a diet focused on stable blood sugar is recommended. At the same time, nutrient deficiencies, like lack of vitamins or protein, can trigger or exacerbate symptoms of depression. A diet containing sufficient proteins, important fats, dietary fibre, and vitamins can also have an extremely positive effect on depression, and at the same time it helps with weight loss.

Green tea also has a supporting effect, both on weight loss and mental-health stabilisation (Kimura et al., 2007).

My conclusion is that excess weight and depression are neither conjoined twins nor two separate worlds. The treatment of one also affects the other, and interventions usually benefit both.

I think it's absolutely right that someone with acute depression shouldn't focus doggedly on the number on the scales. But whatever the case, it does make sense to focus on behaviours like increasing your physical activity and healthy nutrition, and in doing so, to improve both areas together.

Weighing yourself every day is bad for you

The warning that weighing yourself every day isn't a good idea often shows up in the advice offered to people who want to lose weight. The reason is that it's supposedly frustrating and/or obsessive.

Steinberg et al. (2015) published a study on this that showed a connection between daily weighing and higher weight loss. Overweight and obese subjects who participated in a weight-loss program were asked about their weighing habits. The result was that those who went on the scales every day lost about twice as much as the irregular weight-checkers. In addition, the subjects who weighed themselves every day developed an average of 17 weight-control behaviours, while the others developed only seven.

Of course, the subjects who weighed themselves on a daily basis might have been more motivated in general, and their daily weighing was just a consequence of that greater enthusiasm. But even if that were the case, this result debunks the myth that daily weighing somehow reduces the chances of losing weight successfully.

Pacanowski & Levitsky (2015) found significant differences in the results of men and women in their study on the success of the daily weight recording. They tracked people who were losing weight independently; some recorded their weight online each day, in the form of a weight curve, and others did not. For men, daily weight logging apparently had a positive effect, and the male subjects who logged their weight were more successful both in losing weight and later in keeping the weight off. For women, there was no difference between the two groups.

Generally speaking, it's beneficial to your weight loss to monitor your behaviour, be it with regular weighing, food diaries, calorie counting, pedometers, or other methods. The conscious observation of your behaviour alone leads to the desired behaviour. This tendency is echoed by studies that show that people eat less in the presence of mirrors.

For me, personally, daily weighing had another positive aspect, namely, that it was easier to understand the effects of water retention. If my weight shot up by 1 kg in one day, it was clear to me that it had to be down to fluid retention and that it would disappear in the following days.

Because I don't like to start my day by looking at the scales, I got into the habit of weighing myself at night, shortly before going to bed. This also had the effect of putting me off snacking out of boredom close to bedtime, because every bite showed up an hour later on the scales. I knew that 300 g of grapes were not equivalent to 300 g of real extra weight, but on a psychological level, it meant I preferred to do without a snack in order to see a more pleasing number on the scales. Since I've been merely maintaining my weight for months now, that effect has unfortunately dropped off. I still weigh myself in the evening, but the numbers are just a rough check, and no longer represent progress. So I just mentally subtract 500 g if I've recently eaten something heavy, because it makes no difference to me now whether the scales show 64 kg or 64.5 kg.

Being overweight is an illness

The idea that being overweight should be seen as a medical illness was introduced into a new guideline by the American Heart Association in 2013. This led to much heated debate. Experts are still in disagreement. I think describing obesity as a disease is not the right approach.

One of the reasons I believe this is because of the results of psychological research. Different psychological mechanisms have been found that influence our behaviour:

- Self-efficacy expectations, i.e., the belief that we can successfully achieve a certain goal by means of a certain behaviour. Self-efficacy has been found in recent years to be a major factor influencing success in many areas, including educational performance, work, disease management, and athletic achievements. Studies have shown the importance of believing in yourself, especially when it comes to weight loss. Resisting temptation, especially at the start of a diet, is an example of self-efficacy, which Armitage et al. (2014) say provides a good prediction of the dieter's ultimate success. It also works the other way around: in the treatment of anorexia, higher parental self-efficacy in relation to their child's success is a condition for the child's greater success in gaining weight (Byrne et al., 2015). The more we or others believe in our success, the higher our actual chances of success are.

- The influence of stereotypes. Research shows that we do less well when negative stereotypes are activated. For example, Spencer et al. (1999) showed that women performed less well in a maths test when they were reminded immediately before they took the test of the 'women are bad at maths' stereotype, while women who had not been influenced in this way scored the same as the men who took the test.

- Goal setting. Research carried out by Locke & Latham (1990, 2002) shows that the higher the goal, the better people perform. Even if the target seems unattainable, performance is better than with a lower, more achievable target. In spite of the higher achievement, though, the satisfaction it brings is lower. But that doesn't change the fact that more has been achieved.

All these theories can be subsumed under the heading 'self-fulfilling prophecies', which is when our expectations of another person influence them in such a way that they act according to those expectations. In 1968, the psychologist Robert Rosenthal discovered this effect, which was later named after him. He told primary school teachers that some of their pupils had proven themselves in tests to be 'intellectual bloomers', who were expected to show a massive development in their intelligence. In fact, the 'bloomers' were chosen at random. Eight months later, however, precisely those pupils performed much better in a follow-up intelligence test and were also assessed much more positively by their teachers than the other children. This effect on the children's performance was due to the teachers' expectations alone, and so the prophecy that had been made by pure chance was fulfilled (Brehm et al., 2002).

Considering this research, what does our society's current attitude to obesity mean for overweight people? The fat acceptance movement claims weight-loss success rates of less than 5 per cent. Reports about genes, the yo-yo effect, or starvation mode appear in the media, which lead people to think that changing your body weight is an impossible undertaking. The fact that magazines run sensational reports about people who have lost more than 20 kg also reinforces the impression that escaping (severe) obesity is something unbelievable.

When I lost weight, people congratulated me as if I'd achieved an outstanding feat. I can honestly say that I was congratulated more often and more enthusiastically for losing weight than for getting my doctorate.

Perhaps this says something about the emphasis we place on physical appearance in our society, but I also got the impression that some people do see losing 80 kg as a greater achievement than getting a PhD. I used to feel the same way, but that changed completely when I became aware of, and began to question, (my) fat logic. The realisation that I really only have to eat 'normally' and don't have to keep to a particularly complicated or hugely restrictive diet, and that my (excess) weight is not determined by external factors but is completely in my control, ensured that I was able to develop the self-efficacy necessary to make the required changes.

The current attitude in society that obesity as a disease rather than a consequence of behaviour makes things more difficult for overweight people. The position may be well-intentioned, but illness is usually associated in our minds with something beyond our control — something that you get, like a virus or cancer, and something over which we hold no sway. With this idea of overweight people, we create a self-fulfilling prophecy of passivity and helplessness.

A study published in 2015 by Parent & Alquist supports this view. The researchers came to the conclusion that a belief that body weight can't be influenced was associated with unhealthier eating habits and less exercise, which also led to poorer health. The authors refer to previous studies that show that people who read articles about body weight being uncontrollable were less motivated to exercise and were more likely to give up trying to lose weight when they experienced setbacks. The authors point out that the number of books like *Health at Every Size* and media claims about 'set points' and genetic predispositions has recently seen a dramatic increase.

This is one of my reasons for writing this book: I think it's important to offer a counterweight to the prevailing attitude to obesity in our society. Although people believe they are doing good by trying to de-stigmatise being overweight, any suggestion of how difficult it is to change that situation can only make being overweight even more difficult to change. When it comes to the consequences of obesity, the goal should not be to promote a feeling of resignation, but to show possibilities for change, to strengthen self-efficacy, and to create a self-fulfilling prophecy that is positive rather than negative.

I was going to sign up for a Zumba course yesterday, but then Katrin told me she was switching to Mondays, and I'm not going to go alone. It's so annoying. My plans to start exercising never work out.

… he could tell within minutes whether someone was going to stick with it or not. Just by whether they took responsibility for their own fate, or just blamed someone else or the circumstances.

Well, my trainer at the gym told me …

Oh. Okay. How's the training going, anyway?

It's not. How am I expected to train with someone who destroys all my motivation with insensitive comments like that?

A gastric band is a quick and simple solution

Gastric bands and gastric bypasses are now common methods of treatment for severe obesity. A few years ago, I myself had a consultation about having one of those operations. I decided against it at the time. In both procedures, the stomach organ is reduced in size so that only a limited amount of food can be ingested — about the amount contained in a yoghurt pot. A gastric bypass usually also includes removing a section of the intestine, so that any food eaten cannot be digested completely. Both procedures are performed under a general anaesthetic. These carry the usual risks of these kinds of operation — the German Self-Help Organisation for Obesity Surgery quotes a mortality risk of 0.3 per cent for this operation.

The altered digestive system usually leads to severe or less-severe side effects such as digestive problems (especially severe diarrhoea, the so-called dumping syndrome) or problems with the absorption of nutrients.

The belief that obesity surgery is the 'easy option' is very common. It's also the reason why people who have lost weight as a result of such an operation often face similar stigmatisation to overweight people. It's thought that they have 'taken the easy way out', that they are 'lazy' and 'weak-willed' — similar clichés to those associated with being overweight. Carels et al. (2015) found that employers were less likely to hire people who had lost weight through obesity surgery than people who lost weight without the aid of surgery.

Occasionally, people argue paradoxically that obesity can't possibly be cured 'just' by cutting calories, because the illness is so colossal that an operation is often the only thing that can help. The paradox is that they don't make reference to the psychological difficulties associated with maintaining a daily calorie deficit, but to physical difficulties instead. Many people don't seem to realise that the surgery is simply a way to help them eat less.

Surgery merely limits the amount of food that you can take in, and the high success rate, including among people who have 'already tried everything', speaks for itself, or for the effectiveness of calorie reduction. The effects of obesity surgery are no different in principle from those I experienced in the first six months of my weight-loss process, when my intake was restricted to 500 kcal per day. The effect is the same, irrespective of whether you 'just' choose to severely restrict your caloric intake, or you are forced to do so due to a surgical reduction in stomach size.

My advantage, or disadvantage depending on the way you see it, is that I still have the option of stuffing myself. As I now do a lot of sport, I am very happy that I still have the option of eating large amounts (this freedom, of course, includes being able to regain some weight). I also have the advantage of being able to control my nutrient intake better. Deficiencies such as protein insufficiency are common in people who have undergone reduction surgery because the incomplete uptake of nutrients and the diarrhoea often make it difficult to control how much the body actually absorbs from what it is fed. This also means, of course, that some of the calories are not actually consumed by the body and it's easier to achieve a deficit — but at the expense of personal control over your nutrient supply.

All of this shows that surgery is by no means 'the easy way out', but simply a way of supporting people in doing what everyone who wants to lose weight has to do: eat less.

The same principle continues to apply on a physical level, surgery just makes the psychological side easier, because it's not possible to overeat (so easily) any more. But that's not entirely true, either. Approximately 20 per cent of those who undergo surgery fall short of expectations and either don't lose weight (sufficiently) or eventually put on weight again (Crowley et al., 2011).

Surgery by no means guarantees weight loss, despite what many people think. The limitations of the stomach's capacity after surgery can also be 'tricked', and you can still achieve an increase in calories. Sugary drinks, alcohol, and ice cream, for example, are not restricted by a gastric band or bypass, but 'slip right through', and so you can still 'smuggle' an enormous number of calories into your body via a reduced stomach. You can also consume sweets and other high-energy foods (nuts, fats, etc.) in sufficient quantities to lead to weight gain. So even after this kind of operation, it's important to change your eating habits accordingly, because weight loss doesn't come 'easily', or 'automatically', or 'of its own accord' after surgery. A surgical intervention doesn't mean that patients no longer have to put any effort into losing weight.

I'll stick my neck out and say a large part of the success of obesity surgery is thanks to its psychological effect. People are more likely to take advice when they have paid a lot of money for it than if it is free. We tend to defend large investments. Investing in a life-threatening, painful operation can be the trigger for a serious change of diet, so that all that pain you suffered will not have been for nothing. The psychological mechanisms discussed in the previous chapter also contribute to the great success of obesity surgery: by selling it as a guarantee of weight loss, self-efficacy expectations are given enormous impetus, and a powerful self-fulfilling prophecy can develop.

My family, friends, and acquaintances don't think I need to lose weight

This chapter was requested by several people for various reasons. The above statement is, of course, pretty loose, because it's not necessarily related to fat logic — or at least, not that of overweight people themselves. In this case, it's the people around them who succumb to fat logic.

And of course, this statement is only fat-logical when referring to people who are not underweight or for whom losing weight would mean they would become underweight. If an anorexic person expresses the desire to become even more underweight, it's obviously not fat logic when friends and acquaintances react negatively. So this chapter is only about the reactions to someone wanting to reduce their weight to a healthy BMI (above 18.5).

I could probably write an entire book on this subject. Let me start with my own experience. When I weighed 150 kg, there was no one who seriously claimed that losing weight would not be a good idea for me. Apart from my mother, though, as far as I can remember nobody ever asked me about my weight, in all those years. My weight was the elephant in the room, which no one mentioned — until I brought it up myself.

At some point I got into the habit of casually mentioning my weight to signal that I had no problem talking about it. People usually reacted with relief, as if they had just been allowed to see my obesity, after they'd apparently been trying not to notice its existence. In my experience, that's quite a common reaction to overweight people ('Don't acknowledge it, whatever you do!'), but above a certain weight — and 150 kg is definitely above — it becomes farcical.

Doctors never mentioned my weight unprompted, either, which was a relief for me at the time, but which I now view more critically. I don't know if it would have helped to talk about it before I was ready. But it did make it easier for me to convince myself that my weight was not a problem. For a long time, my husband didn't notice that I'd gained weight. Not until I told him myself that I'd put on 20 kg.

Because he saw me every day, it's not implausible that he didn't notice, because my weight gain was very gradual. But he still found me attractive, and I think he also wanted to believe my self-deception about the health impacts of being overweight whenever I told him about the latest articles that showed that obesity wasn't so bad for you. Especially because he had also gained weight over the course of our relationship and was on the verge of class II obesity — and friends or colleagues would often mention *his* big belly and tease him about it.

When I started losing weight, I didn't tell anyone about it. My family lives 600 km away, and I didn't see them until ten months later. I originally intended to keep them completely in the dark, but after six months, I decided that was going too far. So I took my knee operation in April 2014 as an opportunity to tell my family about my intention to lose weight. When we met in September, they thought I'd been losing weight for about four months. My mother had also started losing weight at the beginning of 2014, and thought she was my inspiration. In fact, I think it was the other way around, as I'd already mentioned a few of my new insights when we spoke on the phone ('Do you know that there's no such thing as starvation-mode metabolism? I always thought I'd ruined my metabolism by dieting, but it looks like I was wrong …' or 'Wow, I just read that there's no such thing as a fast or slow metabolism, or at least the differences are apparently very small … I've probably just been eating too much.') At the time, I found it good to talk to my mother about being overweight and losing weight, even though she didn't know much about the process I was going through. She was having very similar experiences, for example, with friends and acquaintances who were 'worried' that she might get 'too thin'.

I was also supported by my husband, who had noticed all my problems with my body and had started to question all his repressed concerns and excuses, just like me.

People often ask me what my husband thinks of my new weight. After all, I weighed about 85 kg when he met me, 150 kg when he married me, and now I weigh 65 kg. That's a lot of ups and downs, and a lot of people can't imagine a partner going along with it.

But a look at his former girlfriends reveals some who were very fat, some very thin, some medium … there have been all sorts. It seems he really doesn't care about body size, but he doesn't want me to get overweight again, especially considering how it affected our lives and my health. It wasn't until months after I'd lost weight that he confessed to me that he, too, had been afraid that I might eventually become dependent on constant medical care.

He was basically the only person who was unreservedly encouraging and supportive. He also lost 15 kg, and he came to the gym with me. I think he really was more than normally wonderful in this respect. While surfing through weight-loss forums, I often read about partners who secretly sabotage their loved one's diets or openly say they don't want them to lose weight. The terrible thing is that that kind of behaviour isn't socially condemned the way it should be.

If someone said that they would find it unattractive if their anorexic partner were to achieve normal weight, they would be guaranteed to be told that theirs was a lousy attitude, and that their partner's health was the most important thing. But when someone says that they prefer 'big' and wouldn't want a partner who was 'all skin and bones', it's usually accepted or even seen as positive. And as soon as someone takes a critical view of the fact that their partner has put on weight, they can expect to be called an arsehole.

Sure, with my weight at 150 kg almost everyone would have understood if my husband had left me. They would have nodded sadly and said my weight really was extreme, so it was understandable. But what if I'd just got a little chubby? Maybe even after having children? He would be expected to accept that — or else expect to be seen as a superficial arsehole.

As far as the broader world beyond my marriage is concerned, I lost my first 40 kg in secret, so to speak, without anyone noticing. When I reached about 105 kg, everyone around me suddenly noticed that I'd lost weight. At over

100 kg, I was still very much within the obese range, but others saw it quite differently. From all sides, I was asked, surely I didn't want to lose any more weight, did I? I must be done with my diet now, right? Yeah, that's terrific losing so much weight, but you don't need to lose any more — surely?

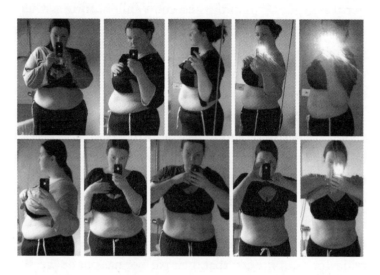

A neighbour who saw me gardening worriedly asked my husband how much I now weighed and asked him to please make sure I ate more. When I ran into a colleague on the street, she half-jokingly asked when I was going to be diagnosed with anorexia, and another colleague admitted that he deliberately hadn't reacted too enthusiastically to my new size, for fear I might go to 'the other extreme'.

The subject of anorexia was brought up more and more by friends and acquaintances. Not that it was directly applied to me, but there were warnings and expressions of concern. There were stories about friends who had also lost a lot of weight and then become anorexic. It was ironic:

when I was sick and almost bedridden at 150 kg, no one had ever expressed concern or commented on my weight in any way. And then, when I had lost 40 kg, I was able to walk again, and I was feeling better than I had for years, people started to get worried about my health? It was as if my body had suddenly become a public forum where everyone was allowed to give an opinion after years of it having been a taboo subject.

Interestingly, it seems this was also only a phase, because now that I'm actually 'thin' at 65 kg, those comments have dropped off a lot. The same people who thought, when I was 100 kg, that I would be too thin and an ugly bag of bones if I lost just one more pound, now either say nothing or tell me I look great.

Psychology can naturally offer an explanation for this, because we always perceive change as being more drastic than stability. When a plane takes off, the speed seems to be faster than it does during flight because we're pressed into the back of the seat by the acceleration. It's probably similar with weight loss, where the weight loss is perceived as more extreme than the stable lower weight after people have had several months to become more familiar with it. Criticism from people around you in this respect is probably something that you just have to endure stoically until people get used to your new appearance.

But that isn't easy to do. Often, people don't limit themselves to harmless criticism or concern, but really become aggressive and censorious. When I started blogging about weight and how society has normalised obesity (I come back to this in the next chapter), I received several emails from people telling me about the response they got to their losing weight. Here are a few examples:

> To be honest, what I found most difficult were the people around me who were constantly trying to convince me that I was anorexic or that I was killing myself 'by degrees' (and that coming from smokers! …), and also the people who kept sending me food parcels out of 'concern'. Some people actually *begged* me to put food in my mouth and were more than a little offended if I didn't eat as much as they wanted (parents and grandparents especially). And of course, the constant know-it-all comments like 'Your body's long since gone into starvation mode — you have to eat more!' or 'Just one meal a day? You can't tell me that your body isn't missing nutrients.' I took comfort in the fact that these people obviously had a different perspective than I did. For me, I can't be called 'skinny' if I can grab enough fat on my hips to really make a handful. I know how much I still want to lose, and I now know that it is perfectly possible. And I also think that a lot of the objections from others come from an 'if-I-can't-have-it-neither-can-you' mentality: 'I wouldn't be able to do it, so you can't do it either. And if you really are getting thinner, then it must be unhealthy in some way. Otherwise, I'd have to come up with a new excuse for not doing it myself.'
>
> I started being more careful about what I ate, and stopped, for example, eating cake just out of a sense

of conviviality with friends. I also refused sweets offered by [my friend] when I wasn't in the mood for them. Then my friend accused me of starving myself and refused to believe that I liked eating the cucumber and kohlrabi sticks I brought to work as a snack. It escalated when she visited me unannounced and 'caught' me leafing through a weight-loss and fitness magazine. 'You're sick! I want my old friend back! Can't you tell that you have an eating disorder? And your obsession with exercise is way over the top!' I was lectured, and then she drove off. After that we didn't talk for a while, but now we're back on speaking terms. But the topics of nutrition, exercise, and body weight are still red rags to her. I find myself hiding fitness magazines from her and hastily making sure no such incriminating publications are around when she visits me. I don't tell her about all the sporting activities I do when we're not together. Although she herself talks about practically nothing other than how many calories there are in certain foods, and tells me regretfully each day what she's eaten, I now completely avoid the subject, because I don't want to be told off for 'being incapable of talking about anything else'. Sometimes, I even lie to her when she delivers her 'diet report' and then looks at me expectantly, and I know that her guilty conscience will not be soothed if I don't admit to eating at least as many calories as her (at the time she accused me of having an eating disorder, one of the things she accused me of was always being careful not to eat a gram more than she did when we were on holiday together. I was very upset by that accusation because I hadn't paid any attention at all to how much she ate. But she was quite adamant, because she had watched closely and counted how many plates I had eaten when we went for sushi …).

I, 1.80 m tall and a proud non-smoker for 11 months, recently got on the scales and weighed 76 kg for the first time. As long as I can remember, my weight has always fluctuated between 69 and 72 kg. Well, there was nothing for it: losing weight was the order of the day. No more sweets, and a few push-ups and crunches until the extra weight is gone (I know, only partly useful for losing weight, but when you lift down the sixpacks from the top shelf, a sixpack is what you want to see).

My wife's reaction was more negative than positive. She liked the fact that I'd put on weight, and now she sees herself almost forced to follow my example (she has gained 17 kg in the last 22 years). Others I've told have been more combative: 'Where have you got spare weight to lose? What am I supposed to say??'

So, it's better to keep quiet about it. I now weigh between 72 and 74 kilos … I'm getting there.

So, after starting to get this feeling of success from comments like 'Oh wow, you've lost 10 kilos! And where did your thighs go?' I've also, kind of unconsciously, started to dress differently. Just dressing more confidently, in clothes I felt made me look nice instead of just less horrible. Before, I used to wear several layers and often long concealing tops (a bit strange, since I was only a little overweight, but we all have our issues^^) but now I've started wearing tighter clothes again. Suddenly, the people around me realised that something had changed. The same people who used to say, 'Oh, nonsense, you're nowhere near fat! You can eat that, no problem!' are now the ones saying, 'Ooh! You've lost a lot of weight! Are you all right? We don't want you wasting away to nothing!' At every. Damned. Opportunity … I've even caught myself putting on extra baggy clothes to avoid these stupid comments when I'm meeting some of the worst offenders. I've also noticed that those people have started giving me bigger pieces of cake at birthdays and putting more food on my plate at dinner parties. (This isn't just in my head — I've kept lists!^^) This is deliberate sabotage!

I could probably show you hundreds of reports like this — and they all follow more or less the same pattern:

- people insist that you are not at all too fat and should not (continue to) lose weight, because then you'll get too thin/ugly or even anorexic/sick;
- people criticise your diet, exercise regime, or target weight as 'too extreme', although it's within the healthy range;
- people urge you to eat certain foods with the argument that you should 'enjoy' eating, or even accuse you of having an eating disorder if you don't partake.

In my case, I was often asked about my target weight, which is in the average normal weight range: 63 kg, with a height of 175 cm, which means I have a BMI of about 21. And although that is a reasonable target, especially considering my damaged joints, everyone, without exception, told me it was 'too low'.

What is shocking is that even my family doctor was uneasy about my target. When I spoke to her while I was still very overweight, she played down the issue and advised me first to try to get somewhere approaching normal weight — that would be great (it sounded more like 'That would be a miracle'). As I got closer and closer to normal weight, she became increasingly critical of my target of 63 kg but was unable to say what exactly bothered her about it, and she had to admit that it was a good target in terms of the risk of osteoarthritis. When I weighed 70 kg, she said that it was enough, now, and that I was looking pretty thin. She asked what my BMI would be at 63 kg — surely in the underweight range? When I replied that it would be 21, she pulled out a little slider calculating chart and eventually muttered that it was true. But she insisted that she still thought it was very low.

If you've read the section in this book on health risks, you will surely have realised that a BMI of less than 21 is associated with the lowest health risk for women. So, in purely medical terms, my doctor should have welcomed my target weight. And she couldn't provide any medical argument against it. Nevertheless, she was visibly, audibly, and tangibly uncomfortable with my aim to become 'so thin'. Her reaction was typical of almost everyone around me.

What I found particularly annoying was that, without even asking, people often assumed, and still do, that I had lost weight purely for reasons of appearance, in order to adhere to 'society's ideal of beauty'. Often, it doesn't help to point out that I actually lost weight for health reasons.

Why do people so often react in such a negative, critical, or hostile way? The people in the comments above mentioned some theories, from the 'if-I-can't-have-it-neither-can-you' mentality (often called the 'crabs in a bucket' mentality) to envy or a guilty conscience. I think, purely intuitively, we all know what the reason is.

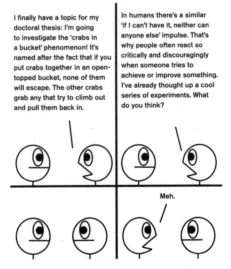

I finally have a topic for my doctoral thesis: I'm going to investigate the 'crabs in a bucket' phenomenon! It's named after the fact that if you put crabs together in an open-topped bucket, none of them will escape. The other crabs grab any that try to climb out and pull them back in.

In humans there's a similar 'if I can't have it, neither can anyone else' impulse. That's why people often react so critically and discouragingly when someone tries to achieve or improve something. I've already thought up a cool series of experiments. What do you think?

Meh.

But the real question, in my opinion, isn't 'Why do they do it?' but 'Why do they think they *can* do it?'

Why is it so socially acceptable to criticise someone for losing weight?

I think this has something to do with the skewed perception of body weight in our society in recent years.

The problem with being overweight is overstated (I don't call it obesity)

This chapter is mostly based on one of my articles titled 'The obsession with thinness and the terror of slimming', in which I wrote about our distorted social perception of weight, and which was commented on and shared quite intensively, provoking some controversy. In it, I showed pictures of women of all weight classes to illustrate how different categories such as 'obese', 'severely obese', 'overweight', 'normal weight', and 'underweight' look, and how our perception has shifted by one category, so that being slightly overweight is perceived as normal, anything below that is seen as (too) thin, and obesity is seen by many as just plumpness.

In our society, the distribution of weight is now divided approximately into thirds: just over a third of people are of normal weight, just over a third are slightly overweight, and just under a third are obese. Only about 1 per cent are underweight. The 'average', i.e., that which is, to a certain extent, the norm, is therefore slightly overweight. In my opinion, this slight excess weight has actually been 'normalised' to such an extent that it is no longer perceived as overweight at all.

Since I don't want to use photos of other people, and I've passed through most BMI categories myself, I use photos of myself to illustrate my point here, even though the effect is probably slightly weaker. If you want to get a better overview of what the normal, overweight, underweight, and obese categories look like, you can also take a look at Kate Harding's BMI project (kateharding.net/bmi-illustrated/), which ironically pursues the opposite goal to mine. She also wants to show how normal, overweight, obesity, etc., look. But her aim is for readers to be amazed at how ridiculous BMI is, when people seem so pretty and average, even though they are obese by category.

So what do we imagine morbid obesity to be? It is the highest possible BMI category, which the fat acceptance movement mockingly likes to call 'Death Fat'. It is the category that is more deadly than smoking, robs people of more than 13 years of life, and marks the upper end of the obesity scale. With a weight of 137 and 145 kg I was well within that range:

145 kg, BMI 47   137 kg, BMI 45

I think this would undoubtedly be perceived as 'fat' or 'very fat'. But do most people imagine morbid obesity to look like this? My impression is that when people think of morbid obesity, they tend to have those 300 kg people in mind, who are sometimes featured in TV documentaries, and who can do nothing but lie in bed and have to be washed with sponges.

This is a picture of me just shy of morbid obesity, with a BMI of 37:

And in this one I am *only* obese:

114 kg, BMI 37

92 kg, BMI 30   92 kg, BMI 30

My body in the last picture is already in danger of all the major health risks of obesity. In my experience, though, a lot of people don't consider someone who looks like this to be obese, but rather chubby or slightly overweight.

In these photos, I am 'slightly overweight'.

The photo with the bike was taken during our holiday, and maybe I weighed a little bit more at the time — about 80 kg — but because of my padded cycling shorts, my bottom and thighs look a bit bigger.

76 kg, BMI 25

... and this picture shows pretty much precisely when I entered the normal weight range — 76 kg with a height of 1.75 m: from this point I could still lose almost another 20 kg before I would enter the underweight category. One kilo more would have made me overweight.

In my experience, at the time when I was still slightly overweight, friends, family, and acquaintances were already of the opinion that I was now normal, and when I had barely reached the upper-normal weight range, I started to get comments like, 'You're already slim, where are you going to lose any more weight from?' The last photo is about what I looked like when my doctor warned me not to lose another 8 kg (ultimately because she wasn't comfortable with the idea).

With a BMI of 23, I was in the upper third of the normal range, whereas the feedback from those around me judged me to be in the lower third and included the opinion that there wasn't much left for me to lose. In the following three

71 kg, BMI 23

photos I finally have an average, normal weight of 65 kg.

In order to reach the lower margin of the normal weight range, I would have to lose about 9 kg at this point. In view of the fact that normal weight is no longer even found in half the population, i.e., it now represents a minority, it's not surprising that the lower end of the normal weight range is already perceived as extreme.

The healthiest BMI from a medical point of view, which has the lowest risk factors for cancer, diabetes, and other chronic diseases, has become the exception and is viewed rather critically. In our eyes, which are so accustomed to slight overweight, it is no longer recognised as normal, but rather as very slim or thin — or at least, those are the words I hear being used to describe me at 65 kg, although I'm almost exactly in the middle of the normal weight range and should actually be described as … well, just normal. When you look at the photo of me on the verge of being overweight, the reaction should be, 'That's a little chubby.'

Some people might react to a comment like that with indignation and say something like, 'Well, that's not what *I* call obese/overweight!', but the categories are defined clearly and don't factor in subjective feelings. Of course, an obese woman can be pretty, and morbidly obese women are also sometimes considered attractive. But that doesn't make them normal weight. Even if they look healthy, their obesity certainly isn't. And our shifted perception won't solve the problem, but will only make it worse. The fact is that the superfluous body fat I'm carrying in the top photo increases my risk of diabetes and cancer, and that's not a matter of opinion, but medical fact.

And yet, when I say this to people, the reaction is often a kind of scepticism, which is partly expressed in the question of whether it might be anorexic to consider the figure shown above as fat. Many people seem to take the attitude 'If I don't *see* it, it's not there.' A 'good (body) feeling' is marketed as being more important

76 kg, BMI 25

than a healthily realistic view. *For the sake of the children, if nothing else! Think of all the kids who are driven to anorexia by super-skinny models!*

But are they really? A new study by Sarafrazi et al. (2014), in which children and teenagers from eight to 15 years of age were questioned about their weight, indicates the opposite. Of the normal-weight children, 87.4 per cent thought they were normal weight. Of the few who did not feel normal weight, twice as many felt they were too thin (8.7 per cent) as felt they were too fat (3.9 per cent).

Among the overweight children, however, fewer than a quarter were aware of the fact that they belonged in that category. Three out of four overweight children considered themselves to be of normal weight. Even among the children who were obese, almost half considered themselves to be of normal weight (41.9 per cent). Only 57 per cent were aware that they were overweight.

Where is the terrible slimming mania that is propagated by the media? Is it not rather the case that, having eyes in their heads, children and young people compare themselves with what they see around them every day? Parents, teachers, friends are much more important. When almost everyone around them is 'carrying a few extra kilos' and, at the same time, there are complaints in the media about models being too thin, although beyond the catwalk film stars and models usually have a completely normal BMI, what are kids likely to consider normal?

This chimes with a recent study by Black et al. (2015) from England, in which parents were asked about the weight of their children. Of the parents of overweight children, 80 per cent rated them as being of normal weight. It was only when the child was already obese that the likelihood of the parents at least thinking of their offspring as being overweight increase. And out of 369 parents of severely overweight children, only four sets of parents stated that their child was *seriously* overweight.

The results of a new US study (Duncan et al., 2015) were even more drastic: 95 per cent of parents of overweight pre-schoolers thought that their child's weight was 'just right', and almost 80 per cent of parents whose child was obese were of the same opinion.

And the situation is no better when it comes to adults. Gannon reports on a 2014 study in the United States, according to which only 36 per cent of Americans consider themselves to be overweight, although 63 per cent of Americans are actually overweight or obese.

In 2012, the *New York Times* reported on a study in which 3662 men and women were asked to determine their body type. Eighty per cent of people of normal weight got it right (but a not inconsiderable number mistakenly classified themselves as underweight), whereas 56 per cent of overweight people thought they were of normal weight. Three-quarters of obese people considered themselves only slightly overweight. The article says diverse studies have produced remarkably similar results, and the researchers theorise that obese people might suffer from a similar body dysmorphic disorder to those suffering from anorexia, but in the opposite direction: while anorexics see themselves as considerably less thin than they really are, overweight people perceive themselves as slimmer.

The article also refers to studies with children in which 70 per cent of overweight and obese children chose a significantly slimmer silhouette than theirs when they were asked to assess themselves on a scale of body shapes. The children with the most overweight parents and friends were out by a greater margin than children with a normal-weight environment.

Paul et al. (2014) added support to those results with their research, which showed that people's own weight strongly influences their perception of others. Thus, fewer than 10 per cent of people of normal weight tended to underestimate the weight of the people around them, but more than 70 per cent of obese people perceived those around them to be slimmer than they really were. That was definitely true for me. When I was still severely obese, I had the feeling that the whole world was skinny, and when I heard talk of an 'obesity epidemic', I would think, 'Where? I'm the only fat person here.'

All this is in line with the idea that our general perception has shifted and that being slightly overweight has already become the norm. And for those who say, 'Yeah, but those are all US studies. That's a long way from the situation here in Germany. Everybody knows the Yanks are fat ...' Well, that view may just be a symptom of the skewed perception that I am talking about, because statistically, we Germans are not so far removed from those 'incredibly fat' Americans.

There is a steady increase in obesity in the USA: while only about 10 per cent of Americans were obese in the 1960s, by the beginning of the 1990s that figure had reached 20 per cent — which is the current level in Germany. In the meantime, the figure has risen to 35 per cent in the USA, and projections say that half of all Americans will be obese by 2030. Since about three-quarters of Americans are already overweight, people of normal weight are likely to form a very small minority by 2030. We are a few years behind the Americans, but the trend is moving in the same direction.

According to a recent government report (Mensink et al., 2013), 1.5 per cent of people in Germany are underweight and 38.5 per cent are of normal weight. That leaves 60 per cent overweight. A total of 23.6 per cent of Germans are obese. So out of five Germans, two are of normal weight, two are overweight, and one is severely overweight. Our average BMI of about 27 points is within the overweight range. Statistics on the actual proportion of overweight people vary widely, depending on whether they're based on telephone interviews, direct surveys, or objective measurements. People tend to misquote their weight to make it seem more normal, which is why overweight rates in surveys are often around 5 per cent lower than in objective measurements (Geiss et al., 2014).

A self-assessment survey by Gallup (2012) on the actual and ideal weight of men and women showed that in the last 20 years men's actual weight has increased by about 7 kg, and that of women has risen by about 8 kg. The ideal weight that respondents would like to have increased by 4.5 kg for men and 4 kg for women over the same period. This increase in desired weight is an interesting development. People are becoming increasingly overweight, but the ideal is also shifting upwards along with their actual weight. What will happen in 20 years' time, when the current generation of overweight and obese children who feel that their weight is normal are questioned?

Incidentally, I was made aware of that research after someone cited it in response to a fairly widespread 'shock' campaign. It aimed to bring the damaging body ideal of extreme thinness to the attention of the public with the headline 'Twenty years ago, the average catwalk model weighed 8 per cent less than the average woman — today it's 23 per cent'. The prosaic answer to this is that models have not discovered a secret trick to getting thinner — the average woman has simply got fatter.

In 2014, *The Guardian* reported on a recent study in which African-American women were presented with a range of different female body types. First of all, they were asked to assess which ones they perceived to be 'too fat'. On average, the women were very much in agreement that this was the case for figures with a BMI of 38 and above — everything below severely obese was 'normal' for them. The women were then asked to assess themselves, i.e., to indicate which of the body types depicted matched their own. Fifty-six per cent of overweight women and 40 per cent of obese women did not see themselves as 'overweight', 'obese', or 'too fat'.

A British study produced similar results. Obese people were asked to assess themselves: only 11 per cent of women and 7 per cent of men with a BMI of over 30 were aware that they were obese.

The researchers then compared the proportion of obese people who considered themselves to be obese or at least 'very overweight' in 2012 with the same proportion in 2007. Fifty per cent of obese people identified themselves as being obese in 2007, compared with 34 per cent who found themselves to be 'very overweight' or 'obese' in 2012. Almost two-thirds of the obese people surveyed considered themselves to be normal, or at least not particularly overweight.

Contrary to the widespread opinion that we have a negative body image, our body image is in fact unrealistically positive, and a large proportion of overweight people don't even realise that they have a problem.

I think the situation is similar to the one a few decades ago, when there was still a widespread belief that smoking wasn't bad for your health, and lots of people blithely took up smoking without knowing what damage they were doing to themselves. According to the US Centers for Disease Control and Prevention, approximately 42.2 per cent of people were smokers in 1965 (and only about 1 to 2 per cent were obese) — until massive campaigns were implemented to point out the dangers of smoking, limit the depiction of smokers in the media, and impose higher taxes on cigarettes.

According to the CDCP, the smoking rate has now dropped to below 20 per cent. The difference with smoking, though, is that it never affected more than half of the population, while obesity already affects the majority of people. Smoking was also understood as a behaviour (choice), whereas many people think that being overweight is a characteristic trait, a 'predisposition' — and therefore something that can't be changed.

The comparison with smoking is also apt on another level, because as with smoking, our social environment has a big influence on our weight. An analysis by Christakis

& Fowler (2007) found the risk of becoming obese increases by up to 171 per cent when a person close to us becomes obese. The more obese our social environment is, the higher the risk is that we will put on weight and thus influence our environment in turn. But this also applies to healthy behaviours: the same study shows that people who adopt healthy behaviours such as not-smoking or regular exercise, also influence their environment, and that those around them have an increased probability of adopting the same healthy behaviour.

The current trend is therefore truly dangerous, as the continuous increase in weight in our society is becoming a vicious circle, leading to an ever-increasing number of overweight people. The more people become aware of this fact and break the cycle, the greater the chance is for those around them to do the same.

Doctors blame everything on weight

You might be wondering why this section is included here among the social issues, rather than in the medical section earlier in the book. I think that this is more of a social problem, because the medical position is clear. There are very few doctors who don't know that obesity is harmful, although I get the impression that a lot of doctors underestimate how harmful it actually is. During all the years I was moving further and further towards a state of severe morbid obesity, I don't ever remember a doctor mentioning my weight. Probably the closest anyone came was to mumble 'your organs are hard to find' during an abdominal ultrasound examination.

When I asked my family doctor a few years ago about the possibility of a stomach reduction, he gave me a referral without hesitating. When I asked him what he thought about it, he said that it would certainly make sense to lose weight if I wanted to have a baby. It would make pregnancy easier. Not a word about the fact that being overweight also had a direct impact on my life, or that I was running a much higher risk of diabetes, a heart attack, joint problems, or premature death.

A recent study by Turner et al. (2014) found that the BMI of only 22 per cent of patients in Britain's National Health Service were documented. In some cases, it hadn't even been recorded for patients with problems directly related to weight. Among diabetics, it was missing in half the cases, and was only documented for 38 per cent of patients with cardiovascular disease or hypertension.

Galuska et al. (1999) asked more than 12,000 obese patients whether their doctor had advised them to lose weight. It was the case for fewer than half (42 per cent).

In an article in *The Sydney Morning Herald* (2015), a doctor complained of precisely this problem from her own practical experience: 'I see plenty of women who come for scans because they have lower limb problems that I know would be eased if they weren't so fat, but no one seems willing to confront them with that.'

This definitely goes both ways, as a lot of doctors have often seen their weight-related interventions ignored or quickly abandoned by patients. Nevertheless, even during my weight-loss process, I found it demotivating to find that almost every medical expert apparently assumed that ultimately I wouldn't succeed in losing weight. When I asked my orthopaedic surgeon after several months on 500 kcal, and having already lost 30 kg, if it was okay with my knee problems to consume less than 500 kcal a day, he just gave an amused snort and said, 'Of course you can do that, but you'll never keep it up!' My impression was that he thought he could tell at first glance that for a woman weighing 120 kg any attempt to make a change was doomed to failure anyway.

As I described in the previous chapter, the only weight-related criticism I have received from a medical professional so far has been about my target weight — which is within the normal weight range.

I don't know how typical my personal experience with doctors is, in this respect. As I've had to move house frequently over the last ten years, I came into contact with many different doctors, but that's still only a tiny selection. There may be doctors who focus more on weight and aren't afraid to discuss it with their patients. And different people might have different impressions about whether too much or too little attention is paid by doctors to body weight.

One reader wrote an interesting review on Amazon:

> The day after reading this book, I went straight to see my lung specialist and told her that she needed to change the weight registered in my medical record. For years, I answered 'no' to the question 'Has your weight changed?' I didn't know what my weight was because I hadn't weighed myself for three years. But my doctor should have noticed that I'd gained 40 kilos in those three years. After all, I see her quite often because I can only get the prescription for my medication from her, and I have to do lung tests regularly as part of a disease management program. I didn't weigh 96 kilos, as was recorded on her computer, but almost 134 kilos. Such a massive change in weight has an immense impact on my medical data, and she should really know that. But I had to ask her directly, twice, to change the weight in my record. She seemed completely overwhelmed by the fact that I was speaking so openly and directly about my weight problem and asked me several times in confusion what it had to do with her. Tests

also showed I had high blood pressure, which she
called 'normal'. Sure, high blood pressure is definitely
'normal' for someone of my weight, but it's certainly
not 'normal' in the sense of being medically healthy.

How many doctors possess the motivation to address
such a potentially unpleasant issue and risk annoying
their patients? Especially when they know from personal
experience that very few patients will actually lose weight
permanently? In this way, a self-fulfilling prophecy is
gradually built up among doctors as well. Even before
broaching the subject of losing weight, they feel it will be
pointless anyway. So, if there is a 'grey area', where weight
is perhaps only one of several causes of ill health, doctors
will prefer to focus on the other causes, in the knowledge
that doing that will get a better reception from patients.

If weight is obviously the cause of an illness, doctors
may address it, but in a rather bashful and casual way
— like when I was told that my organs were difficult to
recognise in an ultrasound exam — and they will hope
that the patient will understand what is meant. When
doctors do address the weight issue directly, it's usually
with a sense of resignation and an attitude that doesn't
motivate the patient. Very few patients then take this as
an opportunity to actually make a change, and the doctors'
resigned attitude is reinforced.

Another problem is that doctors don't live in a vacuum
and so they are also affected by the general shift in
perception of body weight. In a recent study by Robinson
et al. (2014), doctors were asked to assess the photographs

of normal-weight, overweight and obese people. The greater the weight of the person in the photo was, the more likely the doctors were to underestimate it. Overweight people were identified as overweight in fewer than half the cases, and the obese people were recognised as obese in only about one-third of the cases. Two out of three obese people weren't even identified as such by the doctors. At the same time, a physician's perception was found to have a major influence on whether he or she was prepared to address the subject of weight with the patient at all.

I think this development is very alarming, because doctors are still our most important point of contact when it comes to health issues, and even in these times of Google self-diagnoses, the word of a doctor still carries a lot of weight for most people. In the study by Galuska et al. (1999), patients who had been approached by their doctor regarding weight loss were 179 per cent more likely to actually lose weight afterwards. A new study (Bennett et al., 2015) also showed that a good physician–patient relationship was closely related to successful weight loss, especially with regard to weight management. On average, obese patients who experienced their doctor as supportive lost around 50 kg in the two-year study period. The patients who did not evaluate their doctor as helpful, on the other hand, lost only around 2 kg. Of course, perception of the doctor–patient relationship is not a one-way street that depends solely on the doctor, but it does show how great the part that the doctor plays in this process can be.

Doctors can exert a great deal of influence, and I believe they should do so. Patients are affected not only by what doctors say to them, but also by what they don't say. If her doctor doesn't say anything about her weight, a patient interprets it as: 'My doctor doesn't think my weight is a problem.' And if her doctor is motivated to address the issue, the patient can simply dismiss it with, 'My five previous doctors didn't think my weight was a problem. This one here is just incompetent and doesn't want to deal with the real causes, so he'd rather blame everything on my weight.'

You'll slip into anorexia before you know it!

I have often been confronted with anorexia over the past year — or rather, with the word 'anorexia'. Sometimes in the form of a concern, sometimes a reproach, sometimes even as a kind of back-handed compliment. The idea that you can accidentally slip into anorexia seems to be very widespread, as if a diet (or lifestyle change, which is what it ultimately boils down to) were a kind of anorexia banana peel, which healthy people can slip on if they're not careful, and then be sent flying through the air in a wide arc, eventually ending up being fed through a stomach tube.

First of all, a lot of people have a very vague idea of what anorexia is that has nothing at all to do with the medical definition: *Anorexia? That's when you want to get thinner, even though you look totally fine. You get it because you want to be like the skinny models on magazine covers, and you don't understand that they're photoshopped ...*

The medical criteria for anorexia nervosa according to the official International Classification of Diseases (ICD) manual includes that the patient must have an actual body weight at least 15 per cent below the expected weight, or a body mass index of 17.5 or less (for adults). Weight loss is self-induced by avoiding high-calorie food and at least one of the following:

- self-induced vomiting;
- self-induced purgation;
- excessive physical exercise;
- use of appetite suppressants and/or diuretics;
- body dysmorphia in the form a specific mental disorder.

It's important to realise that the criterion of low actual body weight isn't there for nothing. Someone who is overweight or of normal weight cannot have anorexia — at most they may have a different eating disorder. Anyone whose weight is healthy, or even too high, should not be diagnosed as anorexic. This isn't discrimination or ignorance, as I have sometimes heard from the fat acceptance movement. It's simply logical. Perhaps this is best compared to the following situation.

Imagine a person who is extremely frugal. Rather than buying food, she fishes as much as she can out of rubbish bins and dumpsters. She steals toilet paper from supermarket lavatories, and she never turns on the heating, even in the depths of winter, and sits at home wearing two sweaters and a blanket, still freezing.

If this person is a student, with no financial support from the government or her parents, and who has to support herself as best she can with waitressing jobs, never knowing whether she'll have enough money for food at the end of each month, we would probably say her situation was sad, but that her behaviour was understandable.

If the person in question was a bank director earning a six-digit salary, our reaction would be quite different. We would classify her behaviour as some kind of delusional disorder, since there is no objective reason for her extreme thriftiness, and she even runs the risk of causing damage to herself — what if one of her customers sees her stealing rolls of paper from a public lavatory, or fishing food out of a dumpster?

A behaviour should never be considered in isolation. It must be viewed in context, otherwise the image of it will be false. Living on 500 kcal a day for six months was a decision that made sense for me when I weighed 150 kg. If I were to do the same thing weighing 56 kg, thereby forcing my body to use up the fat reserves it needs just to survive, it wouldn't have been a sensible decision, but a harmful one.

It's also not anorexic for a person in the upper or middle range of a healthy BMI to want to be at the lower end of that range. A desire to lose weight must also be seen in context: for an overweight person, it's extremely sensible, for a person of normal weight it's perfectly okay, and for an underweight person it is harmful. There are clear medical criteria here. It doesn't depend on anybody's sense of proportion or what they *think* is normal, too thin, or attractive.

The idea that anorexia is 'the opposite' of being overweight, so to speak, and that overweight people can 'slip' very quickly into anorexia, is also a myth. For one thing, the statistics tell a very different story: while 60 per cent of people are overweight, only 0.3 per cent are anorexic. When it comes to the consequences, the difference is even more extreme. In the US, anorexia accounts for about 0.00673 per cent of deaths annually

(Hewitt et al., 2001), while about 5 to 27 per cent of deaths are because of overeating and overweight (Mokdad et al., 2004; Masters et al., 2013).

While being overweight is already the norm, anorexia is an extremely rare disease. This alone makes it clear that the two phenomena are not based on the same mechanism, and a person is not equally likely to suffer from either condition. Obesity is basically caused by the fact that we follow our instinct to prepare for leaner times too much. Thousands of years of evolution have shaped us to consume as much fatty and sweet food as possible. We experience food as pleasant, rewarding, and positive. Not-eating, on the other hand, runs counter to all our physical instincts, especially when our body weight is at a low.

Although being overweight damages the body and causes illness, our instinct still tells us 'Come on, we still need that piece of cake', because evolution has never developed in such a way as to give the body an upper limit, and so the body never reaches the point where it automatically stops us eating, for our own good. But when it's underweight, our body begins to struggle massively, and the hunger becomes unbearable, even if we might sometimes also experience fasting highs. Still, the body begins to focus exclusively on 'food', our thoughts revolve only around food, we get cravings and feel weak, and our body gives us clear signals that there is a problem. Anorexia is a permanent fight against survival instincts, so to speak.

The underlying problems that cause someone to fight against the physical will to live aren't triggered by a few glamorous photos of skinny models.

Studies have shown, for example, that, unlike bulimia, anorexia is not culture-bound and exists independently of any social standard or ideal (Keel & Klump, 2003). Researchers also found that two-thirds of people with eating disorders reported having experienced sexual abuse in the past and still found it to be the cause of stress in the present (Oppenheimer et al., 1985).

One of my friends worked for several years in a shared-living community for young women with eating disorders and is currently a specialist for eating disorders in a counselling centre. The above findings reflect her practical experience with those women: a high percentage of her clients suffered both from eating disorders and post-traumatic stress disorder after sexual assaults, or they were from highly dysfunctional families.

The idea that anorexia can 'just happen', or that you can slip into it if you're not careful, is a massive trivialisation of a very complex set of problems. As is the belief that phrases like 'Real women have curves, only dogs like to play with bones' might save, or even cure, women and girls affected by anorexia nervosa.

In most cases, the symptoms of anorexia do not include a perception of an extremely underweight ideal as attractive. Rather, people affected usually find a normally slender figure quite attractive in others, but their perception is severely disturbed when it comes to their own body. Although they are underweight, they see themselves as 'fat'. For example, they acknowledge that their slim friends or caregivers have good figures, although they objectively weigh much more than the patients themselves. Treatment of anorexia, therefore, usually focuses on correcting this perceptual disorder. For example, it might include exercises in which patients try on clothes from their normal-weight environment and recognise that they are too big for them.

I recently read in a forum about a man who tricked his anorexic girlfriend into seeing her body differently. He took a photo of her flat stomach, made it anonymous so that she couldn't recognise herself, and asked her to judge the photo. She said that the stomach in the picture was her ideal because it was so beautifully flat and not fat like her own. When her boyfriend told her that she had been looking at her own stomach, she was shocked — but then it became a wake-up call. Although the experience didn't immediately cure her perceptual disorder, it was a first step towards being able to see her own body more objectively.

Contrary to many people's belief, the perceptual disorder behind anorexia nervosa isn't 'fishing for compliments', but rather, the affected person experiences it as enormously traumatic and sees themselves as repulsive. This perception lies outside their conscious control, and well-intentioned comments from people around them — 'But you're not fat!' — are not seen as credible.

The mistaken belief that anorexia is something you choose or pursue like a trendy diet is also often fed by the media. When stars like Meghan Trainor or Jennifer Lawrence talk about how they tried anorexia, but 'couldn't keep it up', the media likes to rip those comments out of context and celebrate them as fat acceptance, even though they're really about a personal fight against the pressure placed in Hollywood on women's bodies and their proportions. If an adult really said, 'I've tried anorexia, discovered that it's stupid, and now I'd rather accept my body as it is!', the person concerned should have it explained to them that at most he or she took an anorexic person as a role model and deliberately tried to starve themselves down to the same weight — and that some of those 'role models' might be unhealthy and thin without actually suffering from anorexia. In a case like that, being convinced that extreme thinness is uncool would actually be enough to cure you of the self-made eating disorder. That doesn't work with real anorexia.

Anorexia nervosa is not a fashion, triggered by thin models and negative attitudes to being overweight. For anorexics and those at risk from the disease, 'interventions' to alter those fashions and attitudes won't affect them in any way.

Instead, anti-anorexia campaigns result in the patients who are affected actually being further split off from society, because it becomes even easier for them to distance themselves. People with anorexia are by no means incapable of thinking, and they recognise the double standards behind positive portrayals of being overweight. They're right to say that health is obviously not the prime concern when such an unhealthy 'ideal' is promoted, and it makes it all the easier for them to rationalise their condition with statements like, 'They're not concerned about my health, they just want me to be as fat as they are.' When you argue with extremes, an extreme stance can be justified by being better than the opposite extreme — be it overweight people who say they'd 'rather be fat than a bag of bones' or anorexic people who say 'at least I'm not a disgusting fatso'. It's much more difficult for someone to justify why a healthy middle way is worse than their own extreme path.

Overweight people already face enough discrimination as it is!

Last year, I occasionally blogged bits of writing and cartoons about being overweight. The topics I covered in my blog were the basis for the contents of this book, and some of the cartoons in this book have appeared previously on my blog. There were two main reactions: some readers wrote to say that they had been encouraged and had finally made a change. Others sent me furious emails, accusing me of discriminating against fat people because I claimed that anyone could be slim. They said that instead I should promote tolerance by pointing out that some people have no choice. Their gist was, 'It's hard enough for fat people as it is, and now you're discriminating against them by saying that they could lose weight if they wanted to.'

I am very much of the opinion that people who are perceived by society as 'fat' face discrimination. This is also supported by experiments such as association tests. In these, subjects' reaction speeds are compared when they are asked to pair positive words such as 'beautiful' or 'happy' or negative words such as 'bad' or 'evil' with silhouettes of either slender or fat people. This can reveal people's unconscious attitudes. Even if they might say, when asked, that they have nothing against overweight people, and they even believe this themselves, they may still react faster to positive words when they are paired with slender figures.

You can find the results of this kind of test on Harvard University's 'Project Implicit' webpage. Around 70 per cent of people who have done their test prefer slim people, 19 per cent have no preference, and only 11 per cent have more positive associations with fat people. This shows that we consider 'slim', socially speaking, to mean more positive.

The *Daily Mail* (2007) published the results of a study showing that the mere sight of overweight people can trigger feelings of disgust and disquiet, similar to the sight of rotten food. The article says this is because, in the course of evolution, our brain has learned to recognise and avoid outward signs of illness, because viruses and bacteria are invisible. People who have a great fear of illnesses also feel the greatest aversion to overweight people.

This sounds tough, in fact it sounds as if overweight people have an extremely hard time in our society.

I have an anecdote on this subject: I knew a member of an online social community who caught my attention because of her extreme hatred of fat people and her constant negative comments about how stupid, lazy, and undisciplined those 'exploded whales' are, and about how disgusting fat people are in general. At some point I came across some pictures of women in bikinis with class I or II obesity that had been commented on by the user in question. I prepared myself for the worst — and then read her irate comments: That it was outrageous to call these 'perhaps a little chubby' women 'fat', when they were just normal women with a few curves.

In my opinion, this anecdote is typical. There are indeed prejudices against fat people, but 'fat' has shifted so far in our social perception that it's usually only the very obese who face discrimination directly.

Of course, there are some people who know that, with a BMI of 28, 30, or 32, they are overweight and feel hurt by such opinions. They read or hear something about 'disgusting exploded whales' and feel like it's aimed at them, knowing that they fall into the category of 'overweight', according to the medical criteria. In fact, they are not the target at all, and the people around them are often astonished when they are offended, because 'You're not fat! I didn't mean you!'

Discovery (2014) confirmed this perception with a report on a study that found weight discrimination to be 'surprisingly rare'. Fewer than 1 per cent of people of normal weight reported experiencing weight-based discrimination. Among the 'slightly overweight', only one in more than 70 people said that they'd received comments about their weight. It was only in the obese group that weight discrimination became more frequent, with an average of 15.6 per cent reporting it, but with a strong focus on the most severe obesity category (35.9 per cent). The researchers also point out that there is a distortion of perception, for example when a star like Miley Cyrus is called 'fat', it's deemed to be newsworthy, while she is described hundreds of times every day (and in different languages) as 'beautiful', 'sexy', or 'hot', and nobody reports on it.

Puhl et al. (2008) came up with similar results, and also separated their figures by gender. Women were found to be much more likely to suffer weight discrimination; especially slightly obese women, who are three times more likely to be discriminated against than slightly obese men. For men, the weight that is perceived as normal appears to be a good deal higher than for women. This is certainly at least partly due to the fact that the proportion of muscle mass in men is biologically higher than in women, and so because muscles are heavier than fat, the same body weight looks slimmer. This study also found that discrimination only became extensive for those with a BMI in the severely obese range.

But what is actually meant by 'discrimination', or 'fat shaming', as it's often called in the US? The definition can vary a lot. Here are some examples that appear in studies:

1. 'Teenagers made "moo" noises.'
2. 'The dentist was worried I might break his chair.'
3. 'Before going to a party, I went to McDonald's so that no one at the party would think I was eating more than I should.'
4. 'My friend's mother refused to give me anything to eat and told me that I'm fat because I'm lazy.'
5. 'My former boss saw me several times in a restaurant and pretended not to know me. I worked for him for five years, and he always hated fat people.'
6. 'I spent the day in the garden, and this survey made me realise how much time I spend alone.'

In fact, only examples 1 and 4 are actually cases of weight discrimination. Numbers 3, 5, and 6 are not actual discriminatory incidents, but rather the interviewees' ideas of what others might have thought about them, and so instead they're more expressions of insecurity or self-doubt. Number 2, on the other hand, is simply reality.

Studies are often quoted that purport to prove that fat shaming just makes fat people fatter. They cite links between the amount of weight gained and the frequency of weight discrimination experienced. This suggests that discrimination causes overweight people to eat more out of frustration.

I think these are just correlations, rather than causes. Looking at the 'incidents of discrimination', you could just as well put forward the theory that people who have more of a problem with being overweight also see discrimination more often (where there may not even be any). It is not surprising that people like that might also be more likely to continue to gain weight: people who identify strongly with being fat are more likely to experience it as an attack when someone mentions their weight, and at the same time they are less likely to have the self-efficacy it takes to lose weight.

The fact that there is a weight limit for certain equipment isn't fat shaming, it's just reality. As I've mentioned, some people have found my cartoons discriminatory, but they're only factual.

We have to make a distinction here between actual discrimination and (partly understandable) oversensitivity. It is never okay to insult or abuse people — for example, by making 'moo' noises. Everyone deserves to be shown basic respect, and it doesn't matter whether or not the person can 'do something about' their condition or not. It doesn't matter to what extent things are within a person's own responsibility or innate, it is not okay to insult or abuse people.

On the other hand, I find the way 'political correctness' in connection with being overweight is developing to be absurd. We have reached the point where statements like 'being overweight is unhealthy' or 'being overweight is caused by overeating' are considered to be discriminatory. In the political fat acceptance movement, even stating your intention to lose weight is considered discrimination, and even more so telling a weight-loss success story. Yes, even the word 'overweight' is discriminatory, as it suggests that people can objectively have 'too much' weight.

Over the past year, I have come to know about people who were inspired by my blogs to do something about their weight. They found it motivating to read the naked truth. Some people said that it hurt them at first, but that was necessary to spur them on to make a real change. Some said that they'd had a guilty feeling in the back of their minds for quite some time, but the people around them made them feel that there was no reason to change, and they were all too willing to believe it.

As we walk ever more cautiously on eggshells in an effort not to offend anyone, important truths can no longer be spoken out loud. Fat shaming may hurt people's feelings in the short term, but fat acceptance kills. Fat acceptance activists like Ragen Chastain are invited in to schools and universities to talk about 'body positivity' and weight. Imagine a cigarette company offering courses at schools about how it's okay to smoke. It is unthinkable that a school would entertain the idea that such an initiative might be a good thing, and if it did, there would be a massive uproar among parents and in the media.

I think we urgently need to change our approach to being overweight. The definition of 'discrimination' or 'fat shaming' should be limited to actual inappropriate behaviour, such as insults, abuse, or interference by strangers that oversteps the boundaries. But when doctors, friends, or family express serious and well-meaning concern, or point out the health risks of obesity and the possibility of losing weight, it is not discriminatory, but caring.

This ideal of extreme thinness is just a modern 'fashion trend'

Marilyn Monroe has somehow managed, posthumously, to become an icon of the fat acceptance movement — even though she found obesity terrible and is on record as saying, 'There is nothing positive about being fat ... Life is more enjoyable when you are thin and pretty. I was never fat a day in my life and I never will.' Ironically, she is often falsely quoted as having said 'To all the girls that think you're fat because you're not a size zero, you're the beautiful one, it's society who's ugly.' Unfortunately, there wasn't a 'size zero' in Marilyn's day.

Marilyn achieved this status as a fat acceptance icon because of a combination of several factors:

- there's a photo of her on a beach in which she is pregnant, and therefore slightly(!) plump;
- she was a sex symbol;
- dress sizes were different in her day and she wore *that era's* size 42.

These factors are combined to create a fantasy that is supposed to confirm that in the 1950s 'overweight' women like Marilyn were considered sex symbols. After all, the average woman today wears that '1950s sex-symbol size'. What more proof do we need that our ideal of beauty has changed radically?

As a matter of fact, Marilyn was about 166 cm tall according to her passport and her dressmaker, and she weighed 53.5 kg, giving her a BMI of 19.5, which is about the same as the BMI of most of today's Hollywood stars. Her body measurements were something like 89 — 56 — 89 in centimetres.

The data and media company Bloomberg carried a report on its website in 2011 about an auction at which one of Marilyn Monroe's dresses was sold. The frock did not fit the model, who was an average dress size of 32, who was hired to wear it for the auction, because the zipper at the waist wouldn't close. That means Monroe would be extremely slim even by today's standards, and I suspect that, with a 56 cm waist, she would be seen today as a 'bad role model' who caused eating disorders in young people by promoting an unrealistic ideal of beauty.

Incidentally, the gradual, unannounced adjustment of dress sizes is known as 'vanity sizing'. Clothing manufacturers know that female customers like to flatter themselves that they can fit into smaller sizes, and that they are more likely to buy a garment if it is marked with a 38 rather than a 40 or 42. *The Washington Post* put together a graphic showing that a US size 8 corresponded to a waist size of about 60 cm in 1958. In 2001 that had already grown to about 69 cm, and in 2011 it was as much as 75 cm. Clothes sizes are therefore generally a poor measure of weight change, as they vary widely between manufacturers on the one hand, and generally increase over the years, on the other.

To return to the ideals of beauty in past decades, other famous Hollywood stars of past eras were also far from overweight, even though they were sometimes described as 'full-bodied', 'voluptuous', or 'curvaceous':

> Elizabeth Taylor (1.60 m/54 kg, BMI 21.1)
>
> Catherine Deneuve (1.68 m/61 kg, BMI 21.6)
>
> Rita Hayworth (1.68 m/53 kg, BMI 18.8)
>
> Raquel Welch (1.68 m/53 kg, BMI 18.8)
>
> Brigitte Bardot (1.70 m/56 kg, BMI 19.4)
>
> Audrey Hepburn (1.70 m/47 kg, BMI 16.3)
>
> Grace Kelly (1.70 m/53 kg, BMI 18.3)
>
> Sophia Loren (1.73 m/64 kg, BMI 21.4)
>
> Ingrid Bergman (1.75 m/63 kg, BMI 20.2)

Audrey Hepburn would probably not even have been allowed on stage today with her severely underweight BMI. Let's compare that with today's sex symbols, who were listed as the sexiest women by *FHM* in 2014:

> 10. Scarlett Johansson (1.60 m/57 kg, BMI 22.3)
>
> 9. Nicole Scherzinger (1.65 m/54 kg, BMI 19.8)
>
> 8. Lucy Mecklenburgh (1.68 m/54 kg, BMI 19.1)
>
> 7. Beyoncé Knowles (1.69 m/61 kg, BMI 21.4)
>
> 6. Mila Kunis (1.63 m/52 kg, BMI 19.6)
>
> 5. Kaley Cuoco (1.70 m/57 kg, BMI 19.7)
>
> 4. Emily Ratajkowski (1.70 m/54 kg, BMI 18.7)
>
> 3. Rihanna (1.73 m/56 kg, BMI 18.7)
>
> 2. Michelle Keegan (1.63 m/54 kg, BMI 20.3)
>
> 1. Jennifer Lawrence (1.75 m/63 kg, BMI 20.6)

There isn't really very much difference between them. What is interesting, though, is that they don't differ much from the stars of earlier decades — and yet I often read and hear from (older) men that they prefer 'full-bodied women' like Raquel Welch or Brigitte Bardot, because they were 'real women', with 'some meat on their bones' unlike the 'skinny girls of today'. I regularly come across similar statements about the 'real women of the good old days' compared to the 'skinny girls of today'.

Is it perhaps because these women were perceived differently at the time? Perhaps they didn't differ so much from the average women of the time in terms of their size, but were just more hourglass-shaped, with a narrower waist and larger breasts, while the average woman was … well, average. So in contrast to 'normal women', these women were seen as 'voluptuous' or 'curvaceous', whereas today the most striking difference is that stars weigh 20 kg less than the average woman, and therefore their most obvious feature is their thinness.

The arguments often go even further back in time, for example, to Rubens's paintings, or other representations of fatter women in art. Now, I am not a scholar of art, but the following things always cross my mind when I hear these arguments:

Who was the subject of the painting? Personal portraits were mostly the preserve of the rich, because average people couldn't afford a portraitist. But in the past, rich people were also the only people who could afford to become overweight. This is certainly the source of some bias.

Even if being overweight was a status symbol, that doesn't necessarily mean it was actually perceived as being the most 'beautiful' condition; it was simply a part of the way people liked to portray themselves. For instance, washing was considered something only peasants did. Aristocrats used perfume to mask their smell and only poor people needed to wash themselves with mere water. Does that mean that the smell of sweat in combination with strong perfume was the ideal scent, and the most pleasant aroma in people's noses?

Many artistic representations of ideal women, such as the Venus de Milo, are absolutely in keeping with our present ideal. The Venus de Milo has visible abdominal muscles, which indicates a very low body-fat value. There's also the question of whether an artist was working to commission, depicting a social ideal, or creating a work that suited his own taste. There are art historians who say Rubens preferred to paint fatter women to show off his painterly skills, because cellulite and rolls of fat are harder to depict realistically than a slender body without any great differences in contouring (Sweet, 2014). Even today, there are very diverse representations of different bodies. That famous photo of Albert Einstein is very popular, but that doesn't mean our ideal of masculinity is an elderly man with a shock of white hair sticking out his tongue. To believe that every representation of a person necessarily reflects the prevailing ideal of beauty is probably also a somewhat single-minded way of thinking about art. Perhaps some painters were also concerned with depicting the decadence of the ruling class by showing their obesity. It's probably possible, at least to some extent, to retroactively reconstruct earlier ideals of beauty on the basis of some paintings, but such a reconstruction will never be a complete one.

Times change. As mentioned earlier, the short-term advantage of a fat supply in the body used to be more important than it is today, because famine could mean death at any time. Keeping sufficient energy reserves in the body was still chronically harmful in the long term, but that was outweighed by the short-term benefit. Today this is no longer the case, because there is little risk in our society of suddenly suffering months of starvation and having to rely on 20 or 30 kg of bodily fat reserves. Why shouldn't our ideal have adapted? What is more, the negative effect of obesity was also observed in the past. The following passages are taken from a guideline for family doctors written in 1854 by Dr George Capron:

> Fatness often becomes a disease. The adipose substance under the skin, in certain instances, becomes as thick as that of a hog ... This pressure narrows the cavity of the chest and produces a difficulty of breathing. The heart and the large blood vessels also become compressed, and consequently the pulse becomes weak and slow ... Fatness produces insensibility and a disposition to sleep, and where it becomes excessive, apoplexy and death. Neither the heart nor the lungs having room to expand, the circulation of the blood languishes, and the motions of the vital organs are oppressed.

… Fatness is commonly induced by high living and a free use of fermented liquors. The laborer, and the active in any pursuit of life, rarely become fat. People in easy circumstances who can enjoy all the comforts of life without sharing in any of its cares and toils, are extremely liable to become corpulent. Care and exercise are sure to make people lean, while quietude and indolence are as sure to make them fat. The means of cure, are therefore in every one's power. There is a wide difference between a fleshy person and a fat person. Too much flesh or muscle can injure no one, but too much fat, which is an increase of the cellular substance, necessarily produces disease.

To reduce the fatness, people only have to reduce the quantity and the richness of their food, and to take such a degree of exercise or to adopt such a system of labor as thinner people pursue.

A newspaper ad from 1905 is similarly direct. In big block-letters, it is addressed 'TO FLESHY PEOPLE' and promises, among other things, 'When you have reduced your flesh to the desired weight, you can retain it. You will not become stout again. Your face and figure will be well shaped, your skin will be clear and handsome, you will feel years younger. Ailments of the heart and other vital organs will be cured. Double chin, heavy abdomen, flabby cheeks and other disagreeable evidences of obesity are remedied speedily.'

Some readers have probably already seen old newspaper advertisements for products that can be used to gain weight easily. A Google image-search for 'vintage ad fat' will bring up countless old newspaper ads, and about half are for weight gain.

The slogans in these adverts are aimed at presenting thinness as unattractive and unhealthy and the ads claim that an extra 5 to 15 pounds will make you more beautiful, healthier, and happier. Such adverts seem inconceivable in our time. But if you look at the 'after' photos, you'll see that the results are still what we would consider to be slim today. The ads above are aimed at 'fat' people. At that time, *that* was considered fat.

And it is by no means true that obesity was celebrated back then. The other half of the ads are aimed at overweight people, telling them that losing weight will make them more beautiful, healthier, and happier.

So the fact that there used to be ads for weight-gain products can't be taken as proof that being overweight was prized. It's just another proof of the shifting of average weight. Then, as now, the healthy normal range was the most attractive size. In times of war, economic crisis, and food shortages, it used to be difficult for some people to avoid being underweight, so there was a market for sweets that made you fatter or other means of gaining weight. Today, when 60 per cent of people are overweight and 1 per cent are underweight, there is no longer a market for such products.

In an article in 2013, *The Spectator* compared the current figures on obesity with those of the late 1960s. While in a survey in 1967 more than 90 per cent of overweight people had taken some action in the previous year to reduce their weight; in a survey conducted in 2010, only just under half of those questioned had done so.

This difference just goes to show clearly, once again, how wrong our present perception of the past is. Being overweight was far less common and was more likely to be seen far more negatively than it is today.

Men like women to have a
bit of meat on them

This chapter also deals with the ideal of beauty, somewhat lop-sidedly focusing on women — but that reflects the way it is in the real world.

I have noticed a strong tendency in contemporary society towards the idea that 'men should find chubby/fat/overweight women attractive', combined with a negative attitude towards the slimmer women and men who admit to being attracted to slim women. Either those men's heterosexuality is called into question (with jokes like, 'Studies have shown that nine out of ten men are attracted to women with curves. The rest are attracted to other men.'), or it's intimated that they might not even like adult women at all, but be drawn sexually to children. The feminist-lesbian initiative *ARGE dicke Weiber* (a heading from their homepage: 'A fat women's working group for physical diversity and positive self-image — fat acceptance in Vienna') writes in its flyer:

> Let's examine the norm we are all supposed to
> aspire to more closely: the dream measurements
> are 90-60-90.
>
> A 90 cm bust measurement corresponds to
> size 38
>
> A 60 cm waist corresponds to children's size

134 cm, which is the average for a child of 8 to 9 years of age

A 90 cm hip measurement corresponds to size 34, or children's size 16.

That means the ideal is a little girl's body with breasts. This ideal is a criminal male paedophile's fantasy.

In a society such as ours, in which the accusation of paedophilia is one of the most serious that can be made, this statement exerts a great deal of pressure. But the depiction of thin women as unsexy or child-like rather than as 'real women' is also highly derogatory and repressive. With my current average, normal weight, my measurements are 90-66-93 — and so it seems I already almost have a 'little girl's body with breasts'.

Meghan Trainor's hit song 'All About That Bass' is also a product of the current popularity of valorising overweight women while simultaneously disparaging slimness. Here are a few relevant quotes from the lyrics of that song:

'... boys they like a little more booty to hold at night ...'

'You know I won't be no stick-figure, silicone Barbie doll ...'

'I'm bringing booty back
Go ahead and tell them skinny bitches Hey
No, I'm just playing, I know you think you're fat,

But I'm here to tell you that,
Every inch of you is perfect ...'

A song like this aims to counter the alleged oppressive emphasis on being thin by instead degrading slim women. This is sold as an anthem for a positive body image. By declaring such degradation of thinner body types to be part of a fight against anorexia, any criticism is nipped in the bud, because the cause is supposedly a good one. In truth, discrimination is never a good thing, whichever direction it goes in, and it can never be helpful or supportive. Unfortunately, those affected are often their own worst enemies, for example, when obese people use this fantasy to protect themselves from having to make changes that are urgently advisable for health reasons.

People who say that overweight people are unattractive are against body acceptance

This is a short, interim chapter, because this is such a common protest. The accusation of being against a positive body image for all and of wanting to undermine (further) the self-confidence of overweight people often comes up when someone makes a critical statement like the one above ('overweight people are unattractive'), or says that fatness is not a general ideal of beauty.

I am absolutely in favour of loving your body and accepting yourself. But I have a completely different understanding of what that means to that of the people who use these terms to actually promote being fat. Accepting and loving a fat body, be it your own or that of a fellow human being, is very human and is definitely nothing to be condemned. But encouraging people to be fat and persuading them that there are no health reasons for losing weight is, in my view, downright reprehensible, because it's nothing except lying to people and making them lie to themselves and to others — in many cases with life-threatening consequences.

If I love someone or something, I want the best for that person or I want to take as much care of the beloved object as I can. We have no problem in pointing out that some things are dangerous and harmful. We tell smokers that smoking is unhealthy. We tell drug addicts that drugs are bad. We insist that people in cars fasten their seatbelts and that children wear bicycle helmets. It's not okay to insult or abuse smokers or drug addicts, but as a society, we have no problem with portraying their habits as bad, unhealthy, and dangerous. A heroin acceptance movement, insisting that arms full of needle pricks and scars are more erotic than healthy arms, would hardly gain many followers. But at the same time, we celebrate it when a mainstream modelling agency gives a contract to a model with a morbidly obese BMI. She's made up, squeezed into a corset, and photoshopped until her figure looks as attractive as possible. This is then presented as a 'body positive' message. How is that different from the hypothetical attempt to boost the self-image of heroin addicts?

Basically, *everyone's* self-confidence deserves to be boosted. But not by presenting unhealthy and harmful behaviours as cool.

Some people seem to think that only perfect people are allowed to be confident in themselves, and therefore you have to act as if everything you do is right and perfect. True self-confidence and self-acceptance, however, means being aware of your own negative behaviours and accepting that you are not an all-round perfect person — and realising that you don't *have* to be perfect in order to be allowed to love yourself.

I think this awareness is also necessary if you are to be able to work on and improve yourself. First of all, you have to accept that there are areas where you are not perfect, rather than ignoring or denying that fact. As far as being overweight is concerned, fat logic is an excellent aid to *not* accepting yourself. If it's the 'fault' of your genes, the environment, your thyroid gland, or anything else, then you don't have to accept that you have a weakness.

In this respect, I am completely and unreservedly in favour of every human being, including and especially overweight people, loving and accepting themselves and being self-confident — just, please, without making it contingent on their 'fat identity', but rather on their person as a whole.

This links in with a study carried out by Carraça et al. (2011), in which obese women were accompanied during a weight-loss program that included work on body image. It turned out to be helpful to improve their overall body image, but, in particular, to reduce its importance in the minds of the participants.

I think this is the crux of the matter: campaigns that focus on a 'positive body image' by trying to present being fat as fabulous and sexy send an implicit message that other people's view of obesity is what's most important. But it is more helpful for people to realise that what other people think about their weight is not the deciding factor. Loving your body and having a positive feeling about it does not depend exclusively on how your body is seen by the people around you and whether they find it beautiful, attractive, or sexy. How it feels to be in that body, and whether you feel comfortable in it, is — or should be — more important. Of course, for most of us, how we are perceived is also important, and it would be unrealistic to dismiss that completely, but our body is not a screen we carry around for others to project their own images onto. Body awareness means awareness of how your body feels to you, not how it feels to have your body judged by others.

For me, my attitude to my own body has changed completely over the last few years. When I weighed 150 kg, and was in pain and unable to move properly, my body often felt like an encumbrance, a problem, a source of worry dragging me down. Body acceptance was not an important issue for me before I lost weight. My feelings about my body were neutral; I neither hated nor loved it. But the fitter I got and the better my body felt, the more I began to perceive my body as a part of me, integrated into myself, with the knowledge that I was dependent on it, and at the same time I felt responsible for it. Real body acceptance and feeling good about my body came after I consciously made and acted on a decision to do something for my body and stop engaging in behaviours that were harmful to it. That's when my body actually began to feel good, rather than feeling like a source of numerous problems, steadily falling apart and in need of constant repairs.

I think it's difficult to feel comfortable in a body that is under strain or in pain. People who are chronically ill, and those with chronic pain in particular, know that this condition can often be accompanied by a real hatred of the body and a feeling that it is the enemy. At least, this has often been an issue in my work with patients suffering from pain, and their treatment usually includes elements aimed at learning to see their body as something positive once again, for example using massages or scented oils.

Loving your own body and feeling good about it is much easier if that body really does feel as good as possible. A real feeling like that doesn't change even if 50 people tell you how good or how bad it looks in their eyes.

After that short digression, back to the topic of whether overweight people can be sexy. The journal *Psychology Today* dedicated a long article to the topic in 2014. The authors report that studies in different cultures all over the world have shown that men always prefer a similar ideal of female beauty, and it can be summed up in one word: curvaceous. I should point out here that although 'curvaceous' has become a euphemism for 'fat', in this context it definitely does not mean overweight. It refers to the classic feminine curves, with a focus on a narrow waist. The ideal waist is 60 to 70 per cent of a woman's hip measurement.

The average Playmate or porn star has a BMI of 18.5, which is precisely at the lower limit of the normal weight range (like most past and present Hollywood stars) and has the average measurements of 89-58-89. The only extreme measurement there is the very narrow waist. The chest and hips are 'normally big'.

On average, then, men tend to prefer 'curves' to extremely thin, underweight catwalk models. While two-thirds of catwalk models (not to be confused with the models shown on TV and posters aimed at the general public) have a BMI of less than 17, only about one in 17 Playmates has such an underweight BMI. And the ten sexiest women chosen by the British men's magazine *FHM* all have a BMI of more than 18.5.

In 2012, Faries & Bartholomew investigated the role of body fat and fat distribution in female attractiveness. Male and female subjects were asked to evaluate body scans of women. On average, both women and men found the figures with the lowest percentage of body fat (15-20 per cent) to be the most attractive, and the higher the body-fat ratio, the less attractive the figures were deemed to be. A narrow waist in relation to wider hips was also preferred by both genders. Wang et al. (2015) conducted a similar study in different cultures and also had men and women in three European, three Asian, and three African countries evaluate body scans. Here, too, the result was that a lower BMI and body-fat percentage was considered to be the most attractive, and a narrow waist was preferred. However, body fat was the most important factor.

So there are two aspects that appear to be decisive in making a woman maximally attractive:

- weight at the lower end of the normal range;
- a narrow waist in comparison to her hips (low waist-to-hip ratio).

That also just happens to be the body type associated with the lowest health risks. It does means that an overweight woman with an hourglass figure can of course be as attractive as a slender woman with a less curvaceous figure, but while bone structure and fat distribution cannot be influenced, and so a woman cannot really influence her basic figure (the most she can do is exercise her underlying muscles, such as building up her butt muscles), she can control her weight as much as she wants.

The argument that all this is purely due to the influence of the visual media is disproved by the fact that blind men also prefer women with narrow waists (Karremans et al., 2010).

Incidentally, when shown images of fully clothed women, men focused most on their face when assessing their attractiveness. When the women's body shape was visible, for example when they were pictured in swimwear, body and face were given about the same weighting (Bleske-Rechek et al., 2014). Furthermore, when the women were fully clothed, their BMI was a better predictor of their attractiveness to the male subjects, but when they were shown in swimwear, their waist-to-hip ratio was the more influential factor.

But the fact that women's faces play a big part in whether they are judged to be attractive is once again not independent of weight. The attractiveness of women's faces has been found to depend heavily on their proportion of fatty tissue (Fisher et al., 2014).

When I lose weight, I'll finally like myself

So, first I spend a whole chapter saying that normal weight is more attractive, and now I'm presenting the nice idea that once you've lost weight, you'll feel more beautiful as an aspect of fat logic? Well … yes, I am.

In fact, I'm convinced that false hopes in this area are quite often a source of disappointment and frustration, and lead to people abandoning their diets. It's not for nothing that I have tried to give priority to the medical and general benefits of having a healthy weight, rather than focusing on appearance. The reason I am writing about it nonetheless, is that being attractive to others is a great motivation for some people, and there's nothing wrong with that.

Personally, I think the health aspect is more important, but that's also because I have already felt the consequences. When I was 20, my priorities were different (which, in my case, was one of the reasons why I then regained weight).

Attractiveness is important in our society, and since you're often considered attractive as the result of signs of good health (such as shiny hair, an upright posture and taut muscles, clear skin, or healthy teeth), the two go hand in hand. So it would be ridiculous and schoolmarmish of me to wag a finger and say, 'Appearances are not important, you have to lose weight for the *right* (i.e., *my*) reasons.'

It's certainly not wrong, shallow, or stupid to hope to get a better feeling about your body, increase your self-esteem, or gain more self-confidence by losing weight. I have often been asked — I think it might even be the question I've been asked most — how my feelings about my own body have changed since losing so much weight.

It's not such an easy question to answer. My appearance was never really important to me. I was always fine to sit down in front of TV and enjoy an episode of *Germany's Next Top Model* with a bar of chocolate and still feel great. I didn't find my 150 kg body disgusting, repulsive, or ugly. When I started to lose weight, my view of my body changed. I looked at things more critically for the first time, consciously perceiving things that I had ignored or suppressed until then: the fact that my stomach hung down over my thighs; and that I had skin folds where there should not be any, and that they rub against each other and gather sweat between them in the summer; and that some areas were unpleasantly lumpy or asymmetrical (the more fatty tissue you have, the more asymmetrically it is distributed — for example, my left knee was much fatter than my right one); and the fact that normal body parts like my collarbones weren't visible at all.

I realised for the first time that our thighs are not naturally designed to rub together when we walk. It was a realisation that really took weeks to sink in properly: thighs are not supposed to rub against each other. It would be totally illogical for our body to chafe in that way while walking normally (people used to walk an average of 35 km a day). For a while after I realised this, I paid attention to people's thighs while out shopping in town, and I noticed that most of them rubbed against each other.

By the way, I'm not talking about the 'thigh gap' — the space between the inner thighs some people have even when they are standing with their feet together. I am talking about normal standing or walking, with your legs hip-width apart. When I reached a weight of about 70 kg, I noticed the lack of friction noise when I was walking for the first time.

When I realised those things, they started bothering me. I was aware of my belly when it lay on my thighs. I pinched at my skin folds or looked critically at my legs in the mirror when they were squeezed together. I still didn't feel bad about it at the time, because I was completely sure that these were temporary features, and whenever I criticised my body, there was always a 'still' at the back of my mind. My legs are *still* rubbing together. My belly is *still* bulging over my thighs. My collarbones aren't visible *yet*. I started sometimes talking negatively about my body, which I had never done before.

This had a lot to do with my altered attitude towards my body (fat). I used to *be* my fat, in a way. 'Fat' was part of my identity and something that made me who I was. When someone criticised fatness, I felt personally criticised. This has now changed, radically. In my mind, my fat turned into no more than an energy store that was too full. I didn't feel like a 'fat person' any more, but rather a person who was 'temporarily still carrying excessive fat'. This is probably down to the fact that my personal fat logic consisted mainly of the belief that I was naturally fat because of my genes and my family background, and being fat was part of my basic makeup. When I started to question that fat logic and began to understand that being fat was merely an (extended) behaviour of mine, I saw it in a completely different way and took a much more objective view.

All in all, this was a very positive development, but it also took on an extreme nature for a short time. I deliberately looked at surgical photos and videos of obese people and the fatty tissue in their abdominal cavity. I looked at fatty livers and fatty hearts, and for a short time, I felt disgusted by the thought of the yellow, bloated tissue that was inside me. Understanding the biological processes and realising the biological purpose of adipose tissue fortunately helped me to regain a more neutral view. I don't know to what extent the process I went through is typical, but plenty of people who have also lost weight, or are in the process of doing so, speak of increased self-criticism and a temporarily less positive body image. I think it's normal to go through a temporary phase of increased dissatisfaction in the course of a process of change, and I also think that it takes this dissatisfaction to push you to make a change at all.

Self-confidence comes mainly through inner factors: pride in achieving certain (partial) objectives, enjoyment of new experiences and the process of change, and also the knowledge that you *can* change things — it does not come so much from the immediate feeling of being more beautiful or more attractive. This can sometimes be disconcerting for the people around you, because they only see a positive change and can't understand why you are now more self-critical or dissatisfied. Self-criticism may even be perceived as offensive (especially if the other person is similarly overweight or even more so) or as a clumsy attempt to fish for compliments.

Another important factor for me was building up muscle. At the age of 20, I lost a lot of weight, but completely without exercising, which is why I was slim, but not at all 'taut'. My body awareness was completely different from what it is today. At that time my bum hurt when I sat down, but I thought that was just the way it was for thin people. Today my bum is padded with muscles, and it doesn't hurt when I'm sitting.

After 30 extremely unsporty years, now, for the first time, I can feel my muscles. I didn't know what it felt like to have muscles. What I used to think of as (defined) muscles when I squeezed my arms and legs was probably mostly fat. I was shocked by how hard muscle mass feels. My body suddenly felt completely different. One night I got up to go to the toilet, half-asleep, and was very taken aback when I sat down and put my hand on my thigh and it felt completely unfamiliar to me. I would often lie in bed in the morning feeling my thighs or ribs and struggling to understand this difference compared to the softness of before. On the one hand, I was proud and pleased, but on the other, it was unfamiliar and felt wrong to me. It took several months for me to 'catch up' with the change.

That might make it sound like I've turned into an extreme body builder, but that's not what I mean. It was just that the contrast to my former self was so stark. With a body-fat ratio of about 19 per cent, I am fit for a woman, but certainly not extremely muscular. But the lack of that centimetre-thick layer of fat covering my entire body was unfamiliar, and it therefore seemed more extreme to me. Seeing and feeling those muscles was a great source of body positivity for me, especially in contrast to my 20-year-old, unfit but slender self. At that time, I was happy not to be fat any more, but my attitude to my own body wasn't particularly improved. This time, the pride and the positive feelings came not so much from my appearance, but from discovering what my body was now capable of and the associated idea that I could be athletic, if I wanted to be. Trying out new sports and being able to do certain movements and exercises successfully straight away, or being able to run several kilometres when I tried jogging for the first time in ten years, gave me much more delight than seeing my slimmer figure in the mirror.

And what about now? Well, I'm in the region of my target weight and have been maintaining it for a few months, and some of those same processes are still ongoing. I'm more aware of my body than I used to be, I look in the mirror more often, and I feel much more positive overall. My skin isn't completely taut in some places, like my bust, belly and thighs. It's not so bad that I would seriously consider surgery, but it's not ideal either. I have set myself various goals — psychologically speaking, it is always easier to achieve a positive goal than to avoid a negative one.

The goal 'I won't put weight back on' would be the worst possible one for maintaining your new weight, because the subconscious mind doesn't understand negative messages. If you say, 'I won't be afraid', your subconscious mind focuses on 'be afraid'. On the other hand, if you say,

'I am going to stay calm', your subconscious focuses on 'stay calm'. And it's the same with your target weight. I don't want to lose any more weight, but that doesn't mean I don't have any more goals. But the goals I have are more like fixed points in the distance that I'm aiming for, and if I don't quite make it, it won't be the end of the world.

A sixpack and tight stomach would be great. To be able to run a half marathon and tackle a grade X climbing route, too. The desire to have a sixpack is a little bit embarrassing for me even as I write this, as I can imagine some readers saying 'How shallow' — but for me it's an incentive, and it gives me motivation. I like to think of Ernestine Shepherd, a body builder who was born in 1936 (I very much recommend a Google image-search) and who, at almost 80 years of age, has the body of a 20-year-old. That is also my goal: to look better and be fitter at 60 or 70 than I was at 20.

Goals like these quickly bring out negative reactions, because it is easy to assume they're born of self-loathing. But I don't hate myself at all. I set myself ambitious goals in other areas of my life, too; but thinking about how I'd like to improve our garden in the spring doesn't mean that I hate the yard. Attending professional further training courses doesn't mean I think I'm bad at my job, and I don't think this book is complete rubbish when I proofread it and add a few lines or correct spelling mistakes.

Goals are motivating. The act of setting goals per se is a sign of self-esteem and self-confidence. I say this as a psychotherapist who sees often enough how difficult it is for patients at the beginning of their treatment to set goals, when they can't even imagine themselves ever getting better.

Goals are great, as long as they're not about achieving performance outcomes, in which the attainment of the goal no longer serves your own benefit, but rather is about gaining a particular (imagined) worth from outside yourself. The difference is that a positive goal is something you want to achieve for yourself — a performance-oriented goal is one that a person hopes will help them *become* something ('I have to graduate with straight A's or I'm a failure' or 'I have to have a thigh gap or I'll never be attractive and never get a boyfriend').

All that doesn't apply to me, thank goodness …

I spent a long time thinking about the best place in the book for this chapter. In the section about the health risks of being overweight? Perhaps. But I consciously decided to put it at the end, since it's aimed at the people who have read this far and (still?) believe that it doesn't apply to them, but to people who are 'really overweight'.

In the country of the blind, the one-eyed man is king, as the saying goes. And in the country where the majority are overweight, the normal-weight man is king. We constantly compare ourselves with those around us, and when the majority of people are extremely out of shape, being 'a bit out of shape' makes us feel almost fit. Okay, maybe you can't spontaneously complete a five-kilometre run, and maybe you get out of breath if the lift breaks down and you have to walk up six flights of stairs, but that's just normal, right? After all, there are some colleagues who wait down in the lobby until the repairman comes to fix the elevator, and others who complain when they have to park in the lot that's a little bit further away. Of course, a few more muscles wouldn't go amiss, and actually, you've been meaning for ages to sign up for that Zumba class … the flyer's still lying around somewhere … you just have to find it and call to ask if there's still room on the course. But Tuesday nights aren't so good for you, as that's usually when you and your colleagues go out for a drink after work, so it would be silly to organise something on those evenings. And, yeah, eating a bit healthier wouldn't be a bad idea, perhaps you should finally give up sugar and white flour for a while — maybe that will boost your energy like everyone says. That would be a good place to a start. But just not until those three packets of muffins and the biscuits in the cupboard at home are finished. As soon as they're gone, you'll buy nothing but healthy food. Definitely.

The term 'skinny fat' has been coined to refer to people who are within the normal weight range but have too much body fat and too little muscle mass. As I already pointed out in the chapter on BMI, around 40 per cent of people of normal weight have proportionally too much body fat. These people run pretty much the same health risks as those who are overweight. Compared to people of normal weight with a normal amount of body fat, their blood has more inflammatory markers, worse blood sugar and fat levels, and they suffer from higher blood pressure (Kang et al., 2014).

Some researchers even believe that 'skinny fat' people have *greater* risk factors than overweight people (which may go some way to explaining the obesity paradox), since their higher body fat levels are not even offset by the increased muscle mass that carrying excess weight around brings. Oliveros et al. (2014) show that 'skinny fat' people who have heart disease have the highest mortality rate compared to all other patterns of adiposity.

The Science of Eating quotes a study that found that one in four people of normal weight are metabolically obese and have prediabetes, full diabetes, or unfavourable blood test results. That is a relatively small proportion, in view of the fact that some studies say 40 to 50 per cent of people of normal weight are 'too fat'.

'Skinny fat' is defined by the following factors:

- looking slim when clothed, but flabby when naked;
- very narrow joints;
- weak arms with little muscle;
- 'muffin top' fat rolls — the belly and bust are the main areas of fat;
- curved shoulders;
- a wide waist;
- a hollow chest.

People who run a high risk of becoming 'skinny fat' include people with eating disorders (due to an insufficient supply of vital nutrients and loss of muscle mass), as well as the elderly, vegetarians who don't eat a balanced diet, people who are sick, and those who do a lot of endurance sport but no weight training. These are all people who either don't use their muscles enough or whose diet doesn't supply them with enough nutrients to 'maintain' them. Männistö et al. (2014) also add smoking and excessive alcohol consumption to the risk factors for becoming 'skinny fat'. As already mentioned in the chapter on smoking, hormonal influences make it more likely for smokers to develop abdominal fat reserves.

The chapter on the health risks of being overweight showed that even people at the upper edge of the normal weight range often have an increased risk for certain medical conditions. This seems certain to be linked to the increasing number of skinny fat people, who are within the normal weight range but have too much body fat in relation to muscle mass. For such people, weight loss is not necessarily important, but reducing fat and building up muscle should be a priority for them if they are to reduce their health risks. 'Skinny fat' is a relatively recent phenomenon, which has only moved into the focus of medical research gradually in the last few years, which limits the number of studies available to me. But it already seems clear that this previously unrecognised group of people run a considerable risk of negative health effects from their weight type.

I've already described my own experiences as a skinny fat person in my twenties and a fit person in my thirties and can only stress once again that they are two completely different worlds. Being 'skinny fat' is not as different from being overweight or obese as people might think. Of course, it's not at all the same thing. Losing weight alone brought enormous benefits for me as far as my physical wellbeing was concerned. I'm just saying it's not the end of the scale. An unfit, thin person can make as big an improvement as an overweight person who loses 15 kg. Both can improve their fitness enormously, and that can have a powerful impact on overall performance.

Unfortunately, our society puts skinny fat people under a lot of pressure to remain that way. 'But you're already thin!' is often the horrified reaction when someone of normal weight decides to work on themselves and change their body. Of course, the reason is that that kind of decision makes other people feel guilty about themselves. They can just about tolerate it when a (severely) overweight person starts to lose weight, since as long as most other people are still less fat than the dieter, that dieter is no threat to anyone's self-image. He really *is* too fat (meaning, fatter than the speaker). But when a thinner person announces that they want to work to improve their body, it can be deemed as tantamount to accusing everyone else of being fat and unfit.

I was particularly struck by this phenomenon when I was browsing through forums looking for 'before and after' pictures. I repeatedly came across people who had started from a relatively normal weight and lost only around 5 kg and built up muscle. The 'after' photos, of course, looked great, but they almost never received any praise from forum readers. More likely were reproachful comments about how the person had already looked good, that they had overdone it, or that they looked more attractive in their 'before' picture. The most praise was garnered by pictures of very severely overweight people who had slimmed down to a weight either at the upper end of the normal range, or in the slightly overweight category.

I think this chapter is especially important because there are a lot of people in this situation who undoubtedly do have a nagging feeling in the back of their minds that they maybe *could* or *should* do something about their condition, but are then persuaded, directly or indirectly (as by comparisons with others), to ignore that nagging feeling.

We can live with the fact that some people we know are naturally slim and sports-crazy, because they have always been that way as long as we've known them. But we can't live with it when our skinny fat friend decides to change — it gives us a guilty conscience, because he's always been as unsporty as the rest of us, 'although he's thin', and he's the person we most like to compare ourselves with.

So skinny fat people are a target group of this book, but at the same time, they're hard to reach. While most (severely) overweight people know that they have a problem, most people of normal weight don't have even an inkling that they might be in a high-risk group. On the contrary, society perceives them as fitter than average. Banishing this kind of fat logic from people's minds is one of the most difficult things to do, because very few skinny fat people will pick up a book like this one of their own accord, as they assume that none of its contents apply to them. That's a shame, because, as I've already said, I can make the comparison from personal experience, and the difference in energy, body awareness, and quality of life between being skinny fat and really fit is enormous.

In summary, reducing body fat makes sense for people who are overweight. But building up muscle mass is just as important, because it brings with it another, huge benefit for both health and quality of life!

Keeping the weight off is the most difficult part

This additional chapter is my solemn farewell to this particular piece of fat logic, which I recently realised had lodged itself in my mind without my even noticing. Sure, I spent a whole chapter explaining that it is not the case that 95 per cent of people are doomed to end up back at their original weight, and I have explained in several ways that there are no magical metabolic changes that mean you will automatically regain lost weight.

Nonetheless, this idea that 'keeping the weight off is the most difficult part' somehow hung over my thoughts like the sword of Damocles. I still remembered how, when I lost that drastic amount of weight when I was 20, I felt like I needed to row frantically, and ultimately futilely, against the tide of regaining weight. I felt uncomfortable whenever anyone mentioned the fact that I had lost weight, and dismissed it with something like, 'Don't speak too soon — keeping it off is the hardest part …' I was constantly worried about what would happen if I regained the weight and everyone criticised me for it. I knew well enough what those conversations were like from the many I had listened to at the family dinner table, when the topic was who had lost weight or who had put it back on. So I tried to limit the damage pre-emptively by quietly and humbly preparing myself for the inevitable yo-yo effect.

This time, free of fat logic and all its erroneous wisdom, I was confident of maintaining my new weight. I never expected it to be easy, but I was sure that my knowledge and my motivation to get my knee back in shape would give me the wherewithal to face the challenge. Now, after a year of maintaining normal weight, I have gradually come to realise that my thoughts about this 'challenge' were actually just another instance of fat logic. Some of the information in the previous chapters was added after they were first written and relates to facts I didn't know about in the first few months of my weight-loss process. For example, the information about the way the chemistry of the brain can make exercise significantly more unpleasant for an overweight person. I discovered my passion for sport gradually over months. As this book makes clear, it all began with weight training. It was a great feeling to be able to watch both my strength and my muscles grow and to see how a body can literally be shaped. That was a completely new experience for me. After learning to love weight training, I gradually also developed a love for endurance exercise, although that really only began as I approached my target weight. Suddenly, my fitness began to improve in leaps and bounds, and I rapidly went from *Yay, I just did twenty minutes on the cross trainer without stopping!* to *What? Has it already been an hour?*

I had several smaller light-bulb moments that had a great impact on me — for example, the first time I reacted to being in a very stressful situation by deciding to take an hour-long bike ride to de-stress. I would never have considered it even remotely possible before. I'd always doubted that other people were telling the truth about this effect (thinking, *Okay, you* say *exercise is relaxing, but … are you sure you're not just trying to convince yourself?*), but even when I did believe that some people might possibly find exercising relaxing, I was sure I could never be one of those people.

Exercise is now part of my daily routine. I have sold my car and now cycle everywhere. Since we live on a mountainside in the country, just shopping or going into town is great exercise. I have integrated my favourite hobby — watching TV series — into my exercise program and now I only watch them while I'm exercising. While I used to be able to easily binge-watch an entire season of my favourite series in a single weekend, the nice side effect is that my new program forces me to draw out the pleasure of devouring the next episode.

All that physical exercise means that my average calorie consumption is now higher than it was when I weighed 150 kg.

That might sound as if everything were perfectly easy now, but that's not necessarily the case. I have to live with the fact that I have a bigger appetite than other ('naturally thin') people. As already touched upon in the chapter on genes etc., there are significant differences in people's appetites. I happen to be one of those people who tend to get hungry after exercising. That means on days when I'm not so active I find it significantly easier to consume *only* around 1800 kcal while on exercise days, even 3000 kcal don't feel like many to me. So, I have to take constant care not to fall into the habit of eating a permanent caloric plus. In principle, nothing has changed from the time when I weighed 150 kg, because, as strange as it may sound, I often had the feeling back then that I was really holding myself back when it came to food. I could always have eaten more than I actually did.

Since I still count calories, however, I am now aware that I actually do eat a lot, so my feeling in that respect has changed considerably. Adapting my eating habits has meant that the amount of food I eat has increased. I still eat sweets if I feel like it, but overall, my entire diet is based far more around healthy foods. Interestingly enough, that hasn't happened as the result of a conscious process of saving calories, but as a consequence of the fact that I developed a better feeling for how what I eat influences the state of my body. When I treat myself to a sumptuous Sunday breakfast of croissants and rolls, of the kind that I used to eat, I'm usually far too stuffed and lethargic to exercise or do anything else. I feel the same way when I eat too much sweet stuff. I never really noticed this effect before, because I was usually lethargic and inactive anyway. Now, I often feel those treats are not worth the unpleasant feeling I know they will cause, and I'm more likely to eat an apple if I'm planning to do some exercise later. Sweets and big meals have now become the exception, and they are restricted mostly to evenings when I don't have any plans for later.

People often ask me if I still constantly count calories. Some readers of my blogs and ebooks have told me that they stopped calorie counting at some point after losing weight and have been able to maintain their weight intuitively since. I'm slightly envious of such people, because that's something I'm still working on. I still count calories, although not as doggedly as I used to. Every now and then I just guess, and sometimes I don't enter an exercise session and the food I ate afterwards into my counter program if I think that they roughly cancel each other out. But when I stop counting completely, I notice how quickly I start to slip back into the classic distortions of perception, and portions start to look far too small, or I forget about things I've eaten. Or I simply eat more. And that's precisely the reason why I, personally, still work with calorie counting. Especially because I am also prone to rather massive fluctuations in retained fluid, of the order of 4 to 5 kg, and so it helps me to know what my caloric balance is. There have been weeks when my weight suddenly shot above 67 kg, only to fall to below 62 kg again a few days later. In such cases, it is very reassuring to know my true caloric balance.

In the long run, I would like to be able to cross over to intuitive eating and no longer have to rely on the aid of calorie counting. On the other hand, I don't find counting calories to be a great chore or tortuous in any way, as it basically takes up only five or ten minutes of my time and it also helps me keep track of my protein intake. I'm willing to sacrifice those few minutes, as counting calories has helped me maintain my weight so well over the past year. If it starts to annoy me some time in the coming months or years and I haven't managed to start eating intuitively by then, I'll probably try out some ideas like intuitive-eating programs or intermittent fasting to see if any of them suit me.

My conclusion after a year is that, for me, maintaining my weight is easier than losing it. For one thing, because movement and activity are now far more pleasant, and for another, because I can now eat more. On a physical level, most people will find the same, but there is also the mental side to consider. And in that context, I can imagine for some people, false expectations like, 'After my diet, I'll be able to eat normally again (normally in the sense of having the same calorie surplus as before),' or hopes of automatic consequences of being thin, can lead to disappointment and make keeping their weight down more difficult. I've dealt with such phenomena in the chapter headed 'If I were thin, all my problems would be solved'.

I have also realised that it was a good thing for me to slim down to my own personal ideal weight, i.e., until my figure was most pleasing to me. There were some times when my weight went down to its lowest level of around 61 kg (BMI 19.9) and, as I now have quite well-developed muscles, my percentage of body fat was low — a little below 18 per cent. Some parts of my body were then a little bit too bony for my taste. The range of 63 to 65 kg turned out to be the one in which I feel most at home in my own body and which I like the most from the appearance point of view. Finding my personal 'feel-good range' really helped me to maintain my weight. When I was 'skinny fat' at the age of 20, my weight fluctuated between 67 and 72 kg, and I was never really satisfied with my body. I think my motivation to maintain that weight was probably less strong because I didn't feel all that much better with my new weight than with the old.

As I see it, as well as finding your own personal 'feel-good' target weight, an important factor in maintaining your weight without stress is to develop an appropriate lifestyle that you find comfortable over the long term and that doesn't leave you feeling that life is a never-ending struggle against being overweight.

Something I encounter all the time is the prophets of yo-yo doom, who tell me to just wait and I'll end up putting weight back on. At first, they said, 'Wait a couple of months,' then, 'Just wait a year,' and now they say, 'Wait five or ten years.' As one commentator wrote, 'When you break your hip in fifty or sixty years' time and suddenly burn far fewer calories because you're no longer so mobile, and you think to yourself, well, I'm ninety years old now and I've had a good life and I'm ready to go, so I won't bother being careful about my caloric intake for the last few weeks or months because it no longer matters, I guarantee someone will wag an arthritic finger at you and a croaky voice will say "See, it's the yo-yo effect! What did I tell you sixty years ago! IT GETS EVERYONE IN THE END!"'

If someone who's having trouble maintaining their weight or who finds it restrictive lets themselves be persuaded that regaining weight is inevitable and the only choice is between 'thin with the feeling of leading a life of restriction' and 'fat but happy', then it really will be difficult to keep the weight off. In effect, this is the kind of thinking that prevents people from looking for the reason *why* they find it so difficult. There are x number of different ways to regulate body weight and if one method is too difficult, there are many others to try out — in principle, it's the same as with weight-loss methods. If someone decides they have to lose weight by cutting out carbohydrates because that's the only true path, but they then find it pure torture doing without fruit and pasta, they might come to the conclusion that losing weight itself is torture, although they might have little problem losing weight with a different method that doesn't involve giving up their beloved spaghetti.

To cut a long story short, this fear from the realm of fat logic escaped the clean-up operation by secretly hiding in a corner, but its time is up and out it goes with the rest.

[Anti-fat logic] Now, I'd still like to know a few tricks!

Most of this is interspersed throughout the previous chapters, but since several people have asked for it, and it's almost obligatory for a psychotherapist writing about weight loss, here are a few psychological aids — or at least possible aids, since they're based on statistics, and every individual must ultimately find out what works best for him or her.

First of all, a basic point: I deliberately avoided going into too much detail in this book about my own weight-loss experience. I have mentioned several times that I restricted my daily caloric intake to 500 kcal at the beginning of the process. This was because of the particular situation I was in. Since I couldn't do any exercise and was under relatively large time pressure, I decided to go for extreme calorie reduction. At the time, I found it astonishingly easy, but it still meant that I ate very little, and my diet was very 'boring'. I mainly ate poultry with a few vegetables, or low-fat curd cheese with protein powder, and of course vitamin and mineral supplements. Anyone *can* do that, but in most cases and for most people it is unnecessarily extreme. For most people, a 1000 or 1200 kcal diet represents a relatively large deficit and results in the loss of about 1 kg per week. At the same time, 1000 kcal allow for a significantly more varied diet than 500 kcal.

In order to make sure I had enough protein in my diet, I was forced to eat almost exclusively low-fat and low-carb foods. Even fruit was banned from my diet at that time for nutritional reasons. If you're at a point where stealing a slice of mango from your husband's fruit salad is a sinful treat, you know you are eating a very restricted diet. When I started gradually increasing my calorie intake, I found eating even 1000 to 1200 kcal very enjoyable.

But there was another reason I didn't go into more detail about my personal diet plan: I think it's important for each individual to find what's best for them, and I don't think much of expounding a certain method and claiming it is suitable for people with very different personalities.

In general, there are several different kinds of diet:

1. Counting. Counting diets include all those which don't prescribe a particular way of eating, but impose a restriction on calorie intake. You can follow this path individually, or you can follow a program like the one run by Weight Watchers. Programs like those simplify calorie counting by translating it into a points system, making it less complicated (but also less precise). The advantage of these kinds of diets is that you can still eat all types of food, so they're very flexible. The disadvantage for some is that they find it tedious to count.

2. Food restriction. This includes those diets which indirectly restrict the amount you eat, by changing the kinds of food you consume. There is no conscious

calorie counting, but the choice of allowable foods increases the likelihood that you will automatically feel less hungry and therefore eat less. These diets mostly restrict the intake of (simple) carbohydrates and/or processed foods, and encourage the eating of either high-protein or high-fibre foods. Low-carb diets like the Dukan, Atkins, or paleo diets fall into this category, as do raw-food diets, whole-food diets, and some vegan diets (although, many sweets and confectioneries are vegan, as is sugar itself, so veganism doesn't necessarily entail a low-calorie diet). When certain foods are forbidden, many people automatically eat less overall. For one thing, the 'permitted' foods are usually more filling, and for another, the effect comes simply from the more restricted choice. The more choice we have, the more we eat, and conversely, a restricted palette of available food types makes us eat less. The advantage of this type of diet is that it allows you to lose weight without feeling restricted in the amount you can eat. The disadvantage for some people is that by 'banning' certain foods, they do feel very restricted in their choices. In addition, for some people, these diets don't 'automatically reduce their appetite'. I, for example, can eat huge amounts of raw vegetables and could gain weight even when eating only fruit and veg. Given that I love cheese, a low-carb diet would be no obstacle to 'achieving' a calorie surplus for me, I'd just guzzle massive amounts of cheese.

3. Eating rules. This includes diets that limit calorie

intake by imposing certain rules. Intermittent fasting is an example of this, in which eating is only allowed during a short time window each day (e.g., between 8.00 am and noon) or regular fasting days are kept (e.g., only water and vegetable broth on Tuesdays and Fridays). In theory, barmy diets like 'food combining' should be included in this category. These diets work by establishing certain rules and structures that amount to food restriction of some kind. The advantage is that a lot of people find it easier to stick to a clear set of rules. The disadvantage is once again that some people can't be 'tricked' in this way. For example, again, for me it would be no problem to eat more than 2000 kcal in the space of four hours, so a time restriction would not automatically make me thinner.

4. Dieting aids. This includes anything that makes it physically more difficult to eat (a lot). In theory, this should also include stomach reduction surgery. 'Stomach fillers' like konjac flour or chia seeds, for example, swell up in the stomach and create the feeling of being full. The advantage here is that some people find they no longer feel restricted in the amount they can eat. The disadvantage is that less extreme measures such as this often don't work and some people will continue to ingest too many calories on top of the filler.

5. Dietary plans. These are, of course, also based on calorie reduction, but they aren't flexible and they are not created by the dieter but are pre-defined.

They may take the form of set shopping lists and recipe collections, deliveries of pre-prepared meals, or a plan drawn up by a dietitian. The advantage is that such plans are usually very well thought-through and include creative and varied meals. They also reduce the amount of planning you have to do yourself, and many people also find it easier to stick to a pre-determined schedule. The disadvantage is that they are imposed from outside and are often very costly — either financially or according to the amount of time they take. Or both.

6. Meal replacement. This type of diet includes all those in which regular meals are replaced in some way. Either in the form of shakes (for example, Slimfast) or by such foodstuffs like juice (juice fasting), cabbage soup, or similar. The advantage is that these diets are relatively simple. The disadvantage is that they usually mean eating becomes boring, rather unpleasant, and monotonous. They can often lead to nutritional deficiencies due to their repetitious nature, and in my experience at least, drinking the same shakes day after day soon makes you so sick of them that the mere thought of drinking another one immediately makes you retch.

7. Increasing the body's calorie consumption. This weight-loss method attempts to create a caloric deficit by increasing activity, without changing eating habits to any significant degree. One hour of intensive(!) exercise each day can easily lead to weight loss of half a kilo a week, while you continue

to eat 'normally'. (However, those who have gained weight continuously should understand that such an exercise program may simply halt their weight gain if they don't also change their eating habits.)

It's impossible to generalise about which type of diet is best. Of course, some forms of dieting fundamentally make more sense than others, or are more likely to lead to success, but there will also be particular types of people who find the best success with their own way of dieting. Personally, I don't have much time at all for juice fasting, as I find drinking calories completely unsatisfying. However, a few weeks ago I read about a woman who lost huge amounts of weight with that method after many other diets had failed. I also don't see the point of spending money on programs like Weight Watchers, when I can do the same thing for free (and more accurately) myself, simply by counting calories. On the other hand, I do understand people who prefer to spend money on gaining access to a simplified points system and also have the opportunity to join a support group.

I think one way of finding out which method suits you best is to analyse the nature of your problem. I would advise everybody to keep a food diary for a week. Ideally, it should be very precise, i.e., with all food weighed out. The information on calorie intake can then form the basis for exploring where the problems might lie. Sometimes this can actually lead to a very simple solution. Someone whose problem turns out to be one of high-calorie snacking between meals might solve it by restricting their meals (e.g., with intermittent fasting) or by changing their eating habits (replacing sugary snacks with raw vegetables). If the problem turns out to be portion size, this could be helped by changing the composition of meal elements (more protein or fibre in the form of vegetables), since this leads to a more rapid feeling of satiety.

Those who have already tried many different methods can use that accumulated experience as a basis for analysing which method worked better and which less well, and then base their diet program on that. Those who have no idea what suits them should try a few different methods out, for example by counting calories for a couple of days, eating a low-carb diet for a few days, trying meal-replacement shakes, or testing a commercial diet plan. Those of a more adventurous bent might want to seek out a number of different programs and then select one at random each morning (by drawing a card or rolling dice).

I think it's absolutely essential to gain at least a basic reading knowledge of nutrition — for example, by looking up the nutritional value of various types of food, calculating your own body's energy and protein requirements, and analysing your own eating behaviour. Not only from the point of view of calorie content, but also from the point of view of a balanced supply of necessary nutrients. Nutrition is one of the most important things in our lives, since it has a direct influence on quality of life. Deficiencies directly affect our ability to function and our long-term physical and mental health. A few hours of work to learn about nutrition (and exercise) might be one of the best investments you can make in your own quality of life.

Everyone who wants to lose weight should find their own path. There is no such thing as the ultimate trick, the one rule, or the one single way of eating that you have to follow. The important thing is to find the best way *for you* to implement the basic, simple idea (fewer calories in than out).

Below is a list of a few minor, basic pointers that might be of help:

1. 'The more people I tell about my diet, the more likely I am to stick to it' is a myth.

2. Positive fantasy and reality. Apparently, the most motivating combination is one of positive fantasy and hard reality (Oettingen, 1996). Hard reality in the form of regular weighing or measuring,

depending on personal preference, is not difficult to implement. My motivational fantasy was my first visit to see my family — after almost a year of losing weight without them knowing. I imagined how shocked and happy my mother would be. Incidentally, the reality was in fact almost as I had imagined it. The last time my mother had seen me, I'd weighed 150 kg, and when she saw me again, my weight was almost exactly half that. She stared at me, completely speechless, for about two minutes and then was totally overjoyed. That fantasy gave me a lot of motivation, and at the same time, I weighed myself every day and followed the curve on the graph to work out what my weight might be by the time I was due to see my mother.

3. Concrete resolutions. For me (and not only me), a successful strategy for 'keeping it up' was to make clearly formed resolutions. Instead of saying to myself *I will do more exercise* or *I'll go jogging more often*, I made resolves like *I will do a quarter of an hour's upper body training every day before I watch the evening news* or *Every Tuesday and Thursday after the team meeting at work I will go straight to the gym*. Attaching the resolution to a concrete external trigger takes the decision away from your brain, so to speak, making it more likely to happen automatically, and less likely that you will start vacillating about whether you should do it or not, so you'll just do it.

4. Stimulus control. This sounds simple but it's extremely effective: simply do not keep temptations

within reach, and instead, place things you want to do within your view. Not keeping sweets at home, making it necessary to go out and buy them especially, creates a considerably higher hurdle to overcome. And keeping your cross trainer constantly at the ready in front of the television, rather than covered in clothes in the corner of your bedroom, will make overcoming unwillingness to exercise that bit easier. I found home-made popcorn works very well as a substitute for 'sweets' (if, like me, you eat it fat and sugar free, it's actually quite healthy and full of dietary fibre), because it takes some effort to make. In addition, researchers have found out that self-control is a limited resource. This means that if you've already had to resist something several times, you're more likely to give in the next time, because your 'power of resistance' has run out (Muraven et al., 1998). So it makes sense to limit the temptations that surround you.

5. Prepare alternatives. It can be helpful to prepare a list of possible activities, in advance, that you can turn to as alternatives to (over)eating (e.g., if you are stressed the list could include taking a bath, reading a good book, watching your favourite series, going for a walk, or relaxation exercises like meditation or yoga).

6. Keep an eye on your nutrient balance. You should have regular blood tests and, if necessary, use supplements, and make sure you are getting enough protein. A deficiency can quickly lead to

loss of energy, making you feel constantly tired and leaving you without any reserves for other activities. Nutrient deficiencies can also cause fluid retention, which leads to frustration on the scales.

7. Protein. Protein-rich meals make you feel fuller, and so it can be easier to deal with cravings if you can find high-protein alternatives to normal foods. Claims that a protein-rich diet places increased demands on the kidneys of healthy people have been disproved numerous times, including, for example, by Manninen (2004), who undertook a comprehensive analysis of previous studies and found no evidence of kidney or any other organ damage in people who ate twice to three times as much protein as the recommended amount. On the contrary, a high-protein diet was found to be associated with lower blood pressure and a long-term positive effect on bone density.

8. A basic weight-training program. This can be done completely for free using YouTube videos, other instructional materials, or even just very basic exercises like squats and push-ups. The exercises should focus on building muscle mass, so when an exercise becomes too easy and eight to 15 repetitions are no problem to complete, they should be made harder with (extra) weights rather than by adding more repetitions. So, rather than doing 300 squats, it is better to do three sets of 12 squats while holding a 10 kg weight. It is also important to make sure you don't exercise the same muscles every day. You should alternate training days and rest days for each muscle

set. It's during the rest days that muscles grow, and too much training can even inhibit muscle development.

9. 15-minute walks. Several studies, for example those by Oh & Taylor (2012) or Ledochowski et al. (2015), have shown that 15 minutes of brisk walking is highly effective in reducing cravings for chocolate and other sweets. A 15-minute walk is not only effective for countering acute snack-attacks, but also as a preventive measure. It was found that subjects had fewer cravings for sweet things in the hours after a walk.

10. Tetris. Tetris? Yes, Tetris. Skorka-Brown et al. (2014) found in their study that playing Tetris for a few minutes reduced cravings for sweet things. The researchers believe this is due to an effect called 'elaborate intrusion', which means that concentrating on a visual task distracts the brain so that it cannot conjure up images of sweets. If you don't like Tetris, any other visually-based game or activity will presumably work just as well.

11. Small, additional exercise routines. Short but intense efforts that quickly increase the heart rate are ideal. If you don't have time for full exercise sessions, you can gain a lot from five to ten minutes of intense step-climbing or even jogging on the spot.

12. Green tea. Quite apart from its weight-loss effects, green tea is very good for your health. It reduces blood pressure, is good for the heart, helps relieve stress and depression, strengthens the immune system, and gives you energy. It's not a miracle cure

for obesity, but this tea can help improve the basic conditions of your body.

13. Water. This is the only negative-calorie foodstuff, which means it doesn't burn infinite numbers of calories, but it is good for your health and helps prevent thirst signals from being misinterpreted as hunger signals. Apparently, a 'mix up' of this kind can occur because the hypothalamus (part of the brain) is responsible for regulating both appetite and thirst. Dehydration can trigger signals which are then erroneously interpreted as hunger. Alissa Rumsey, former spokesperson for the American Academy of Nutrition and Dietetics, says it is therefore a good idea to drink a glass of water right after you get up in the morning, and also to drink some water and then wait a few minutes whenever you have a craving. Drinking a lot can also prevent fluid retention.

14. Every calorie counter needs a set of kitchen scales (available for a couple of euros) and a calorie-counting program/app. MyFitnessPal.com is one such calorie-counting program.

15. Weigh yourself regularly. When and how you weigh yourself is a matter of personal preference. The only thing that's important is to stick to a routine, because your weight can fluctuate a lot throughout the day. Personally, I like to weigh myself before I go to bed at night because I don't want to start the day by looking at the scales, and the thought that I am about to get on the scales often prevents me from snacking in the evening. I prefer to weigh

myself every day, particularly because my weight can fluctuate a lot from one day to the next. If I only weighed myself once a week and then happened to do so on a day when my weight spiked, that would bother me more than weighing myself every day and knowing that yesterday I was 1 kg lighter and so the spike must be because of fluid retention.

16. Habits. Low body weight is associated with stable eating habits: in their daily lives, slim people and those who have lost a lot of weight almost always follow eating routines. This is echoed by studies that show that too much choice of food leads to overeating. Of course, that doesn't mean that you have to live off low-fat cottage cheese and lettuce alone, but it does make a lot of sense to seek out certain foods that you like and can eat regularly.

17. Habits again. Studies like the one by Neal et al. (2011) show that we eat too much out of habit, even when we don't particularly like the food we are eating. Habitual popcorn eaters at a cinema ate stale popcorn, although they would have rejected it in any other environment. The researchers found a similar effect of eating too much when subjects ate with their dominant rather than their non-dominant hand. When we break our habits, we become more conscious of what we eat, enjoy it more, and take in fewer calories. That means it's a good idea to change your surroundings often, eat with your non-dominant hand, lay the dinner table differently, or make any other changes to your eating habits that you can think of.

18. Emotional support. The people around you might not always be supportive, so it can help to speak to people who are going through a similar process to you and have a similar view of it. Personally, I got a lot out of this Reddit site www.reddit.com/r/fatlogic. That forum inspired the title of this book and provided me with a lot of motivation in several ways. In real life, too, it can help to seek support elsewhere if your friends and acquaintances can't provide it. So, go out and look for gym buddies — who are ideally already fitter than you are.

That final pointer 18 was also one of the reasons why I thought it absolutely imperative to write this book, and why in it I don't only concentrate on the physical and mental mechanisms behind weight loss, but also deal extensively with the social aspects and their implications. You will probably find yourself in situations in which the people around you are not just unsupportive, but actively trying to sabotage your efforts or put obstacles in your way. One of the aims of this book is to help you recognise myths and manipulation ('You're not eating enough to lose weight!'). Another of the book's aims is to offer support when the people around you are not providing it. I hope that it succeeds in these aims, and that within its pages you have found the tools necessary to conquer fat logic.

Bibliography

ABC News (2013) 'Early puberty tops list of surprising obesity effects'. Online at: abcnews.go.com/Health/obesity-tied-early-puberty/story?id=20788754#3

Adams, K.F., Schatzkin, A., Harris, T.B., Kipnis, V., Mouw, T., Ballard-Barbash, R., Hollenbeck, A., & Leitzmann, M.F. (2006) 'Overweight, obesity, and mortality in a large prospective cohort of persons 50 to 71 years old'. *New England Journal of Medicine.* 355(8), 763–78.

Al-Adsani, H., Hoffer, L.J., & Silva, J.E. (2013) 'Resting energy expenditure is sensitive to small dose changes in patients on chronic thyroid hormone replacement'. *The Journal of Clinical Endocrinology & Metabolism.* 82(4), 1118–25.

Alpert, S.S. (2005) 'A limit on the energy transfer rate from the human fat store in hypophagia'. *Journal of Theoretical Biology.* 233(1), 1–13.

Amatruda, J., Statt, M., & Welle, S. (1993) 'Total and resting energy expenditure in obese women reduced to ideal body weight'. *Journal of Clinical Investigation.* 92, 1236–42.

American Heart Association (2013) '2013 AHA/ACC/TOS guideline for the management of overweight and obesity in adults'. *Circulation.* Online at: circ.ahajournals.org/content/129/25_suppl_2/S102

Anderson, J.W., Konz, E.C., Frederich, R.C., & Wood, C.L. (2001) 'Long-term weight-loss maintenance: a meta-analysis of US studies'. *American Journal of Clinical Nutrition.* 74(5), 579–84.

Appleton, S.L., Seaborn, C.J., Visvanathan, R., Hill, C.L., Gill, T.K., Taylor, A.W., & Adams, R.J. (2013) 'Diabetes and cardiovascular disease outcomes in the metabolically healthy obese phenotype: a cohort study'. *Diabetes Care.* 36(8), 2388–94.

Archer, E., Hand, G.A., & Blair, S.N. (2013a) 'Validity of US nutritional surveillance: national health and nutrition examination survey caloric energy intake data, 1971–2010'. *PLOS One.* 8(10): DOI: 10.1371/annotation/c313df3a-52bd-4cbe-af14-6676480d1a43

Archer, E., Hand, G.A., Hébert, J.R., Lau, E.Y., Wang, X., Shook, R.P., Fayad, R., Lavie, C.J., & Blair, S.N. (2013b) 'Validation of a novel protocol for calculating estimated energy requirements and average daily physical activity ratio for the US population: 2005–2006'. *Mayo Clinic Proceedings.* 88(12), 1398–407.

Armitage, C.J., Wright, C.L., Parfitt, G., Pegington, M., Donnelly, L.S., & Harvie, M.N. (2014) 'Self-efficacy for temptations is a better predictor of weight loss than motivation and global self-efficacy: evidence from two prospective studies among overweight/obese women at high risk of breast cancer'. *Patient Education and Counseling.* 95(2), 254–58.

Arnold, M., Pandeya, N., Byrnes, G., Renehan, A.G., Stevens, G.A., Ezzati, M., Ferlay, J., Miranda, J., Romieu, I., Dikshit, R., Forman, D., & Soerjomataram, I. (2014) 'Global burden of cancer attributable to high body-mass index in 2012: a population-based study'. *The Lancet Oncology.* Early Online Publication. 26 November 2014

Aune, D., Saugstad, O.D., Henriksen, T., & Tonstad, S. (2014). 'Maternal body mass index and the risk of fetal death, stillbirth, and infant death: a systematic review and meta-analysis'. *Jama.* 311(15), 1536–46.

BBC News. (2014) 'Skirt size increase linked to breast cancer risk, says study'. Online at: www.bbc.com/news/health-29351249

BBC News. (2015) 'Parents rarely spot child obesity'. Online at: www.bbc.com/news/health-32069699

Bell, J.A., Hamer, M., Sabia, S., Singh-Manoux, A., Batty, G.D., & Kivimaki, M. (2015) 'The natural course of healthy obesity over 20 years'. *Journal of the American College of Cardiology.* 65(1), 101–02.

Bellisle, F. & Drewnowski, A. (2007) 'Intense sweeteners, energy intake and the control of body weight'. *European Journal of Clinical Nutrition.* 61, 691–700.

Bellisle, F, McDevitt, R., & Prentice, A.M. (1997) 'Meal frequency and energy balance'. *The British Journal of Nutrition.* 77(1), 57–70.

Bennett, W.L., Wang, N., Gudzune, K.A., Dalcin, A.T., Bleich, S.N., Appel, L.J., & Clark, J.M. (2015) 'Satisfaction with primary care provider involvement is associated with greater weight loss: results from the practice-based POWER trial'. *Patient Education and Counseling.* 98(9), 1099–105.

Berman, M.G., Kross, E., Krpan, K.M., Askren, M.K., Burson, A., Deldin, P.J., Kaplan, S., Sherdell, L., Gotlib, I.H., & Jonides, J. (2012) 'Interacting with nature improves cognition and affect for individuals with depression'. *Journal of Affective Disorders.* 140(3), 300–05.

Berrington de Gonzalez, A., Hartge, P., Cerhan, J.R., Flint, A.J., Hannan, L., MacInnis, R.J., Moore, S.C., Tobias, G.S., Anton-Culver, H., Beane Freeman, L., Beeson, W.L., Clipp, S.L., English, D.R., Folsom, A.R., Freedman, D.M., Giles, G., Hakansson, N., Henderson, K.D., Hoffman-Bolton, J., Hoppin, J.A., Koenig, K.L., Lee, I.M., Linet, M.S., Park, Y., Pocobelli, G., Schatzkin, A., Sesso, H.D., Weiderpass, E., Willcox, B.J., Wolk, A., Zeleniuch-Jacquotte, A., Willett, W.C., & Thun, M.J. (2010) 'Body-mass index and mortality among 1.46 million white adults'. *New England Journal of Medicine.* 363, 2211–19.

BFR (2014) *Bewertung von Süßstoffen und Zuckeraustauschstoffen.* Online at: www.bfr.bund.de/cm/343/bewertung_von_suessstoffen.pdf

Black, J.A., Park, M., Gregson, J., Falconer, C.L., White, B., Kessel, A.S., Saxena, S., Viner, R.M., & Kinra, S. (2015) 'Child obesity cut-offs as derived from parental perceptions: cross-sectional questionnaire'. *British Journal of General Practice.* Online at: bjgp.org/content/65/633/e234

Black, R.N., Spence, M., McMahon, R.O., Cuskelly, G.J., Ennis, C.N., McCance, D.R., Young, I.S., Bett, P.M., & Hunter, S.J. (2006) 'Effect of eucaloric high- and low-sucrose diets with identical macronutrient profile on insulin resistance and vascular risk — a randomized controlled trial'. *Diabetes.* 55(12), 3566–72.

Bleske-Rechek, A., Kolb, C.M., Steffes Stern, A., Quigley, K., & Nelson, L.A. (2014) 'Face and body: independent predictors of women's attractiveness'. *Archives of Sexual Behavior.* 43(7), 1355–65.

Bloomberg. (2011) 'Myth of zaftig Marilyn: Virginia Postrel'. Online at: www.bloomberg.com/view/articles/2011-06-24/hollywood-auction-ends-myth-of-zaftig-marilyn-virginia-postrel

BMJ. (2014) 'Higher sugar intake linked to raised risk of cardiovascular mortality, study finds'. Online at: www.bmj.com/content/348/bmj.g1352

Bralic, I., Tahirovic, H., Matanic, D., Vrdoljak, O., Stojanović-Špehar, S., Kovacic, V., & Blažeković-Milakovic, S. (2012) 'Association of early menarche age and overweight/obesity'. *Journal of Pediatric Endocrinology and Metabolism.* 25, 57–62.

Brehm, S.S., Kassin, S.M., & Fein, S. (2002) *Social Psychology.* Fifth Edition. Boston: Houghton Mifflin Company.

Burton, S., Creyer, E.H., Kees, J., & Huggins, K. (2006) 'Attacking the obesity epidemic: the potential health benefits of providing nutrition information in restaurants'. *Journal of Public Health.* 96 (9), 1669–75.

Byrne, C.E., Accurso, E.C., Arnow, K.D., Lock, J., & Le Grange, D. (2015) 'An exploratory examination of patient and parental self-efficacy as predictors of weight gain in adolescents with anorexia nervosa'. *International Journal of Eating Disorders.* doi: 10.1002/eat.22376

Calle, E.E., Thun, M.J., Petrelli, J.M., Rodriguez, C., & Heath, C.W. (1999) 'Body-mass index and mortality in a prospective cohort of US adults'. *New England Journal of Medicine.* 341, 1097–105.

Cameron, J.D., Cyr, M.J., & Doucet, E. (2010) 'Increased meal frequency does not promote greater weight loss in subjects who were prescribed an 8-week equi-energetic energy-restricted diet'. *The British Journal of Nutrition.* 103(8), 1098–101.

Capron, G. & Slack, D.B. (1854) *Popular Medicine, or, The American Family Physician.* New York: Ray & Brother.

Carels, R.A., Rossi, J., Borushok, J., Taylor, M.B., Kiefner-Burmeister, A., Cross, N., Hinman, N., & Burmeister, J.M. (2015) 'Changes in weight bias and perceived employability following weight loss and gain'. *Obesity Surgery.* 25(3), 568–70.

Carpenter, K.M., Hasin, D.S., Allison, D.B., & Faith, M.S. (2000) 'Relationships between obesity and DSM-IV major depressive disorder, suicide ideation, and suicide attempts: results from a general population study'. *American Journal of Public Health.* 90(2), 251–57.

Carraça, E.V., Silva, M.N., Markland, D., Vieira, P.N., Minderico, C.S., Sardinha, L.B., & Teixeira, P.J. (2011) 'Body image change and improved eating self-regulation in a weight management intervention in women'. *International Journal of Behavioral Nutrition and Physical Activity.* 8(75).

Carrillo, A.E. & Flouris, A.D. (2011) 'Caloric restriction and longevity: effects of reduced body temperature'. *Ageing Research Reviews.* 10(1), 153–62.

CBC News. (2015) 'Diet research built on a "house of cards"? Nutrition studies depend on people telling the truth. But they don't'. Online at: www.cbc.ca/news/health/diet-research-built-on-a-house-of-cards-1.2968704

Centers for Disease Control and Prevention (2012) 'Heart health concerns for NFL players'. Online at: www.cdc.gov/niosh/pgms/worknotify/pdfs/NFL_Notification_01-508.pdf

Chaddock-Heyman, L., Hillman, C.H., Cohen, N.J., & Kramer, A.F. (2014) 'III. The importance of physical activity and aerobic fitness for cognitive control and memory in children'. *Monographs of the Society for Research in Child Development*. 79(4), 25–50.

Champagne, C.M., Bray, G.A., Kurtz, A.A., Monteiro, J.B., Tucker, E., Volaufova, J., & Delany, J.P. (2002) 'Energy intake and energy expenditure: a controlled study comparing dietitians and non-dietitians'. *Journal of the American Dietetic Association*. 102(10),1428–32.

Chang, Y., Kim, B., Yun, K.E., Cho, J., Zhang, Y., Rampal, S., Zhao, D., Jung, H., Choi, Y., Ahn, J., Lima, J.A.C., Shin, H., Guallar, E., & Ryu, S. (2014) 'Metabolically-healthy obesity and coronary artery calcification'. *Journal of the American College of Cardiology*. 63(24), 2679–86.

Chastein, R. (2013) 'My big fat finished marathon'. Online at: danceswithfat.wordpress.com/2013/12/03/my-big-fat-finished-marathon/

Chavarro, J.E., Toth, T.L., Wright, D.L., Meeker, J.D., & Hauser, R. (2010) 'Body mass index in relation to semen quality, sperm DNA integrity, and serum reproductive hormone levels among men attending an infertility clinic'. *Fertility and Sterility*. 93(7), 2222–31.

Chiolero, A., Faeh, D., Paccaud, F., & Cornuz, J. (2008) 'Consequences of smoking for body weight, body fat distribution, and insulin resistance'. *American Journal of Clinical Nutrition*. 87(4), 801–09.

Chirinos, J.A., Gurubhagavatula, I., Teff, K., Rader, D.J., Wadden, T.A., Townsend, R., Foster, G.D., Maislin, G., Saif, H., Broderick, P., Chittams, J., Hanlon, A.L., & Pack, A.I. (2014) 'CPAP, weight loss, or both for obstructive sleep apnea'. *New England Journal of Medicine*. 370, 2265–75.

Choi, H.K., Atkinson, K., Karlson, E.W., & Curhan, G. (2005) 'Obesity, weight change, hypertension, diuretic use, and risk of gout in men:

the health professionals follow-up study'. *Archives of Internal Medicine.* 165(7), 742–48.

Christakis, N.A. & Fowler, J.H. (2007) 'The spread of obesity in a large social network over 32 years'. *The New England Journal of Medicine.* (357), 370–79.

Church, T.S., Thomas, D.M., Tudor-Locke, C., Katzmarzyk, P.T., Earnest, C.P., et al. (2011) 'Trends over 5 decades in US occupation-related physical activity and their associations with obesity'. *PLOS One.* 6(5): e19657.

Citymarathon (2015) 'Deutschland macht (kein) Sport'. Online at: citymarathon.blog.de/2009/05/25/deutschland-macht-sport-6175492/

Clark, A.M., Thornley, B., Tomlinson, L., Galletley, C., & Norman, R.J. (1998) 'Weight loss in obese infertile women results in improvement in reproductive outcome for all forms of fertility treatment'. *Human Reproduction.* 13(6), 1502–05.

Clarke, P., Engelaer, F., & Bauman, A. (2012) 'Olympians live longer than the general population … but cyclists have no survival advantage over golfers'. *BMJ.* Online at: www.bmj.com/press-releases/2012/12/13/olympians-live-longer-general-population-cyclists-have-no-survival-advanta

Cleveland Clinic (2013) 'Obesity: a risk factor for Alzheimer's'. Online at: health.clevelandclinic.org/obesity-a-risk-factor-for-alzheimers/

CNN. (2010) 'Twinkie diet helps nutrition professor lose 27 pounds'. Online at: edition.cnn.com/2010/HEALTH/11/08/twinkie.diet.professor/

CNN. (2012) 'Weight loss may reverse artery clogging, study suggests'. Online at: edition.cnn.com/2010/HEALTH/03/01/weight.loss.reverse.artery.clogs/

Colditz, G.A., Willett, W.C., Rotnitzky, A., & Manson, J.E. (1995) 'Weight gain as a risk factor for clinical diabetes mellitus in women'. *Annals of Internal Medicine.* 122(7), 481–86.

Crocker, M.K., Stern, E.A., Sedaka, N.M., Shomaker, L.B., Brady, S.M., Ali, A.H., Shawker, T.H., Hubbard, V.S., & Yanovski, J.A. (2014) 'Sexual dimorphisms in the associations of BMI and body fat with indices of pubertal development in girls and boys'. *The Journal of Clinical Endocrinology & Metabolism.* 99(8), 1519–29.

Crowley, N., Borckardt, J., Budak, A.R., Byrne, T.K., & Thomas, S. (2011) 'Predicting weight loss success one year after gastric bypass surgery using the inventory of binge eating situations'. Online at: www.muschealth. com/nutrition/documents/Distiguished%20Dietitian%20Award%20 Entries/GPposter%20abstract.pdf

Cuddy, A. (2012) 'Your body language shapes who you are'. *TED Talks.* Online at: https://www.youtube.com/watch?v=Ks-_Mh1QhMc

Cummings, J.M. & Rodning, C.B. (2000) 'Urinary stress incontinence among obese women: review of pathophysiology therapy'. *International Urogynecology Journal.* 11(1), 41–44.

Dagfinn, A., Saugstad, O.D., Henriksen, T., & Tonstad, S. (2014) 'Maternal body mass index and the risk of fetal death, stillbirth, and infant death: a systematic review and meta-analysis'. *Journal of the American Medical Association.* 311(15), 1536–46.

Daily Mail. (2007) 'Scientists discover why thin people dislike fat people'. Online at: web.archive.org/web/20090321045533/ www.dailymail. co.uk/news/article-471711/Scientists-discover-people-dislike-fat-people.html

Daily Mail. (2015) 'Are you a drinkorexic? Rising number of women are swapping food calories for alcoholic ones, experts warn'. Online at: www.dailymail.co.uk/health/article-2947890/Are-drinkorexic-Rising-number-women-swapping-food-calories-alcoholic-ones-experts-warn. html

Daily Mail. (2015) 'Lean meat, papaya and apples: The ten guilt-free foods that burn MORE calories than they contain'. Online at: www.dailymail. co.uk/femail/food/article-2961282/The-ten-guilt-free-foods-burn-calories-contain.html

Daily Mail. (2015) 'Mother whose daughter died after taking diet pills she bought online reveals she had been bullied at school for being ginger and pleads for others "not to take the nasty drug"'. Online at: www.dailymail. co.uk/news/article-3048862/Mother-daughter-died-buying-diet-pills-online-makes-impassioned-plea-not-nasty-drug.html

da Luz, F.Q., Hay, P., Gibson, A.A., Touyz, S.W., Swinbourne, J.M., Roekenes, J.A., & Sainsbury, A. (2015) 'Does severe dietary energy restriction increase binge eating in overweight or obese individuals? A systematic review'. *Obesity Reviews.* 16(8), 652–65.

De Laet, C., Kanis, J.A., Odén, A., Johanson, H., Johnell, O., Delmas, P., Eisman, J.A., Kroger, H., Fujiwara, S., Garnero, P., McCloskey, E.V., Mellstrom, D., Melton 3rd, L.J., Meunier, P.J., Pols, H.A.P., Reeve, J., Silman, A., & Tenenhouse, A. (2005) 'Body mass index as a predictor of fracture risk: a meta-analysis'. *Osteoporosis International.* 16(11), 1330–38.

Dhabuwala, A., Cannan, R.J., & Stubbs, R.S. (2000) 'Improvement in co-morbidities following weight loss from gastric bypass surgery'. *Obesity Surgery.* 10(5), 428–35.

Diabetes Prevention Program Research Group (2009) '10-year follow-up of diabetes incidence and weight loss in the Diabetes Prevention Program Outcomes Study'. *The Lancet.* 374(9702) 1677–86.

Discovery (2014) 'Weight discrimination is surprisingly rare, study finds'. Online at: news.discovery.com/human/psychology/weight-discrimination-is-surprisingly-rare-study-finds-141007.htm

Discover Magazine. (2006) 'DNA is not destiny: the new science of epigenetics: discoveries in epigenetics are rewriting the rules of disease, heredity, and identity'. Online at: discovermagazine.com/2006/nov/cover

Dixon, J.B., Dixon, M.E., & O'Brien, P.E. (2003) 'Depression in association with severe obesity changes with weight loss'. *Archives of Internal Medicine.* 163(17), 2058–65.

Dobrosielski, D.A., Patil, S., Schwartz, A.R., Bandeen-Roche, K., & Stewart, K.J. (2014) 'Effects of exercise and weight loss in older adults with obstructive sleep apnea'. *Medicine and Science in Sports and Exercise.* 47(1), 20–26.

Dolgin, E. (2009) 'Harvard prof falsified sleep data'. *The Scientist.* Online at: www.the-scientist.com/?articles.view/articleNo/27316/title/Harvard-prof-falsified-sleep-data/

Donahoo, W.T., Levine, J.A., & Melanson, E.L. (2004) 'Variability in energy expenditure and its components'. *Current Opinion in Clinical Nutrition and Metabolic Care.* 7(6), 599–605.

Donnelly, L. (2015) 'Obesity poised to overtake smoking as key cause of cancer'. *The Telegraph.* Online at: www.telegraph.co.uk/news/health/

news/11640373/Obesity-poised-to-overtake-smoking-as-key-cause-of-cancer.html

Doyne, E.J., Ossip-Klein, D.J., Bowman, E.D., Osborn, K.M., McDougall-Wilson, I.B., & Neimeyer, R.A. (1987) 'Running versus weight lifting in the treatment of depression'. *Journal of Consulting and Clinical Psychology*. Vol. 55(5), 748–754.

Dulloo, A.G., Duret, C., Rohrer, D., Girardier, L., Mensi, N., Fathi, M., Chantre, P., & Vandermander, J. (1999) 'Efficacy of a green tea extract rich in catechin polyphenols and caffeine in increasing 24-h energy expenditure and fat oxidation in humans'. *American Journal of Clinical Nutrition*. 70(6), 1040–45.

Duncan, D.T., Hansen, A.R., Wang, W., Yan, F., & Zhang J. (2015) 'Change in misperception of child's body weight among parents of American preschool children'. *Childhood Obesity*. 11(4): 384–93. doi:10.1089/chi.2014.0104.

Eckel, R.H. & Krauss, R.M. (1998) 'American Heart Association call to action: obesity as a major risk factor for coronary heart disease'. *Circulation*. 97, 2099–100.

Eneli, I.U., Skybo, T., & Camargo, C.A. (2008) 'Weight loss and asthma: a systematic review'. *Thorax*. 63, 671–76.

Erlinger, S. (2000) 'Gallstones in obesity and weight loss.' *European Journal of Gastroenterology & Hepatology*. 12(12), 1347–52.

Evening Standard (2012) 'Almost one third of London men are too fat to see their own genitals, study finds'. Online at: www.standard.co.uk/news/health/almost-one-third-of-london-men-are-too-fat-to-see-their-own-genitals-study-finds-8191932.html

Fabricatore, A.N., Wadden, T.A., Higginbotham, A.J., Faulconbridge, L.F., Nguyen, A.M., Heymsfield, S.B., & Faith, M.S. (2011) 'Intentional weight loss and changes in symptoms of depression: a systematic review and meta-analysis'. *International Journal of Obesity*. 35, 1363–76.

Faries, M.D. & Bartholomew, J.B. (2012) 'The role of body fat in female attractiveness'. *Evolution and Human Behaviour*. 33, 672–81.

FEELguide. (2015) 'New research discovers that depression is an allergic reaction to inflammation'. Online at: www.feelguide.com/2015/01/06/

new-research-discovers-tha-depression-is-an-allergic-reaction-to-inflammation/

Fernandes, M.F., Matthys, D., Hryhorczuk, C., Sharma, S., Mogra, S., Alquier, T., & Fulton, S. (2015) 'Leptin suppresses the rewarding effects of running via STAT3 signaling in dopamine neurons'. *Cell Metabolism.* DOI: dx.doi.org/10.1016/j.cmet.2015.08.003

Ferrada, P., Anand, R.J., Malhotra, A., & Aboutanos, M. (2014) 'Obesity does not increase mortality after emergency surgery'. *Journal of Obesity.* Vol. 2014, Article ID 492127, 3 pages, 2014. DOI:10.1155/2014/492127

Finlayson, G., Caudwell, P., Gibbons, C., Hopkins, M., King, N., & Blundell, J. (2011) 'Low fat loss response after medium-term supervised exercise in obese is associated with exercise-induced increase in food reward'. *Journal of Obesity.* Online at: www.ncbi.nlm.nih.gov/pmc/articles/PMC2945657/

Fisher, C. I., Hahn, A.C., DeBruine, L.M., & Jones, B.C. (2014) 'Integrating shape cues of adiposity and color information when judging facial health and attractiveness'. *Perception.* 43(6), 499–508.

Flegal, K.M., Kit, B.K., Orpana, H., & Graubard, B. I. (2013) 'Association of all-cause mortality with overweight and obesity using standard body mass index categories: a systematic review and meta-analysis'. *Journal of the American Medical Association* 309(1), 71–82.

Gallup (2012) 'Americans continue to adjust their ideal weight upward'. Online at: news.gallup.com/poll/158921/americans-continue-adjust-ideal-weight-upward.aspx

Galuska, D.A., Will, J.C., Serdula, M.K., & Ford, E.S. (1999) 'Are health care professionals advising obese patients to lose weight?' *Journal of the American Medical Association.* 282(16), 1576–78.

Gannon, M. (2014) 'What, me fat? Most Americans don't think they're overweight, poll finds'. *Live Science.* Online at: bodyrecomposition.com/muscle-gain/muscle-gain-math.html/

Geier, A.B., Rozin, P., & Doros, G. (2006). 'Unit bias. A new heuristic that helps explain the effect of portion size on food intake'. *Psychological Science.* 17, 521–25.

Geiss, L.S., Wang, J., Cheng, Y.J., Thompson, T.J., Barker, L., Li, Y., Albright,

A.L., & Gregg, E.W. (2014) 'Prevalence and incidence trends for diagnosed diabetes among adults aged 20 to 79 years, United States, 1980–2012'. *JAMA*. 312(12), 1218–26.

Golay, A., Allaz, A.F., Ybarra, J., Bianchi, P., Saraiva, S., Mensi, N., Gomis, R., & de Tonnac, N. (2000) 'Similar weight loss with low-energy food combining or balanced diets'. *International Journal of Obesity and Related Metabolic Disorders*. 24(4), 492–96.

Golay, A., Eigenheer, C., Morel, Y., Kujawski, P., Lehmann, T., & de Tonnac, N. (1996) 'Weight-loss with low or high carbohydrate diet?' *International Journal of Obesity and Related Metabolic Disorders*. 20(12), 1067–72.

Goldstein, D.J. (1992) 'Beneficial health effects of modest weight loss'. *International Journal of Obesity and Related Metabolic Disorders: Journal of the International Association for the Study of Obesity*. 16(6), 397–415.

Grodstein, M., Goldman, M.B., & Cramer, D.W. (1994) 'Body mass index and ovulatory infertility'. *Epidemiology*. 5(2), 247–50.

Grover, S.A., Kaouache, M., Rempel, P., Joseph, L., Dawes, M., Lau, D.C., Lowensteyn, I., (2015) 'Years of life lost and healthy life-years lost from diabetes and cardiovascular disease in overweight and obese people: a modelling study'. *Lancet Diabetes Endocrinology*. 3(2), 114–22.

The Guardian (2014) 'Do you know what too fat looks like?' Online at: www.theguardian.com/society/the-shape-we-are-in-blog/2014/sep/10/obesity-body-image

The Guardian (2014) 'Obese Britons don't think they have a weight problem, report finds'. Online at: www.theguardian.com/society/2014/nov/14/obese-britons-dont-think-they-have-weight-problem

Guh, D.P., Zhang, W., Bansback, N., Amarsi, Z., Birmingham, C.L., & Anis, A.H. (2009) 'The incidence of co-morbidities related to obesity and overweight: a systematic review and meta-analysis'. *BMC Public Health*. 9(88). DOI: 10.1186/1471-2458-9-88

Gudbergsen, H. (2012) 'Weight loss is effective for symptomatic relief in obese subjects with knee osteoarthritis independently of joint damage severity assessed by high-field MRI and radiography'. *Osteoarthritis and Cartilage*. 20(6), 495–502.

Gunstad, J., Strain, G., Devlin, M.J., Wing, R., Cohen, R.A., Paul, R.H.,

Crosby, R.D., & Mitchell, J.E. (2010) 'Improved memory function 12 weeks after bariatric surgery'. *Surgery for Obesity and Related Diseases.* 7(4), 465–72.

Guyenet, S. (2014) 'Why do we overeat? A neurobiological perspective'. Lecture online at: www.youtube.com/ watch?v=Mp2p4TdLn_8

Hakala, K., Stenius-Aarniala, B., & Sovijärvi, A. (2000) 'Effects of weight loss on peak flow variability, airways obstruction, and lung volumes in obese patients with asthma'. *Chest.* 118(5), 1315–21.

Hammoud, A.O., Wilde, N., Gibson, M., Parks, A., Carrell, D.T., & Meikle, A.W. (2008) 'Male obesity and alteration in sperm parameters'. *Fertility and Sterility.* 90(6), 2222–25.

Hari, J. (2015) 'The likely cause of addiction has been discovered, and it is not what you think'. *Huffington Post.* Online at: www.huffingtonpost.com/ johann-hari/the-real-cause-of-addicti_b_6506936.html

Hauner, H., Bramlage, P., Lösch, C., Schunkert, H., Wasem, J., Jöckel, K., & Moebus, S. (2008) 'Übergewicht, Adipositas und erhöhter Taillenumfang. Regionale Prävalenzunterschiede in der hausärztlichen Versorgung'. *Deutsches Ärzteblatt.* 105(48), 827-33; DOI: 10.3238/ arztebl.2008.0827

Haupt, A., Thamer, C., Staiger, H., Tschritter, O., Kirchhoff, K. & Machicao, F. (2009). 'Variation in the FTO gene influences food intake but not energy expenditure'. *Experimental and Clinical Endocrinology & Diabetes.* 117(4), 194–97.

HealthDay (2013) 'As years spent obese rise, so do heart risks'. Online at: consumer.healthday.com/vitamins-and-nutritional-information-27/ obesity-health-news-505/as-years-spent-obese-rise-so-do-heart-risks-678335.html

Herz, A. (1997) 'Endogenous opioid systems and alcohol addiction'. *Psychopharmacology.* 129(2), 99–111.

Hewitt, P.L., Coren, S., & Steel, G.D. (2001) 'Death from anorexia nervosa: age span and sex differences'. *Aging & Mental Health.* 5(1), 41–6.

Hofstetter, A., Schutz, Y., Jéquier, E., & Wahren, J. (1986) 'Increased 24-hour energy expenditure in cigarette smokers'. *New England Journal of Medicine.* 314(2), 79–82.

Högström, G., Nordström, A., & Nordström, P. (2014) 'High aerobic fitness in late adolescence is associated with a reduced risk of myocardial infarction later in life: a nationwide cohort study in men'. Online at: dx.doi.org/10.1093/eurheartj/eht527

Hollands, G.J., Shemilt, I., Marteau, T.M., Jebb, S.A., Lewis, H.B., Wei, Y., Higgins, J., & Ogilvie, D. (2015) 'Portion, package or tableware size for changing selection and consumption of food, alcohol and tobacco'. *Wiley Online Library*. Online at: cochranelibrary-wiley.com/doi/10.1002/14651858.CD011045.pub2/full#pdf-section

Hoshi, A. & Inaba, Y. (1995) [Risk factors for mortality and mortality rate of sumo wrestlers]. *Japanese Journal of Hygiene*. 50(3), 730–36.

Hurst, R.T., Burke, R.F., Wissner, E., Roberts, A., Kendall,C.B., Lester, S.J., Somers, J., Goldman, M.E., Wu, Q., & Khandheria, B. (2010) 'Incidence of subclinical atherosclerosis as a marker of cardiovascular risk in retired professional football players'. *The American Journal of Cardiology*. 105(8), 1107–11.

Janowitz, D., Wittfeld, K., Terock, J., Freyberger, H.J., Hegenscheid, K., Völzke, H., Habes, M., Hosten,N., Friedrich, N., Nauck, M., Domanska, G., & Grabe, H.J. (2015) 'Association between waist circumference and gray matter volume in 2344 individuals from two adult community-based samples'. *NeuroImage*. 122(15), 149–57.

Jebb, S.A., Goldberg, G.R., Coward, W.A., Murgatroyd, P.R., & Prentice, A.M. (1991) 'Effects of weight cycling caused by intermittent dieting on metabolic rate and body composition in obese women'. *International Journal of Obesity*. 15(5), 367–74.

Jeffrey, R.W., Wing, R.R., & French, S.A. (1992) 'Weight cycling and cardiovascular risk factors in obese men and women'. *American Journal of Clinical Nutrition*. 55(3), 641–44.

Jeukendrup, A.E. (2011) 'Nutrition for endurance sports: marathon, triathlon, and road cycling'. *Journal of Sports Sciences*. 29:suppl 1, S91–9, DOI: 10.1080/02640414.2011.610348

Johansson, L., Solvoll, K., Bjørneboe, G.E., & Drevon, C.A. (1998) 'Under- and overreporting of energy intake related to weight status and lifestyle in a nationwide sample'. *American Journal of Clinical Nutrition*. 68(2), 266–74.

Jonsson, S., Hedblad, B., Engström, G., Nilsson, P., Berglund, G., & Janzon, L. (2002) 'Influence of obesity on cardiovascular risk: twenty-three-year follow-up of 22,025 men from an urban Swedish population'. *International Journal of Obesity and Related Metabolic Disorders: Journal of the International Association for the Study of Obesity.* 26(8), 1046–53.

Johnston, B.C., Kanters, S., Bandayrel, K., Wu, P., Naji, F., Siemieniuk, R.A., Ball, G.D.C., Busse, J.W., Thorlund, K., Guyatt, G., Jansen, J.P., & Mills, E.J. (2014) 'Comparison of weight loss among named diet programs in overweight and obese adults a meta-analysis'. *Journal of the American Medical Association.* 312(9), 923–33.

Johnstone, A.M., Murison, S.D., Duncan, J.S., Rance, K.A., Speakman, J.R., & Koh, Y. (2005). 'Factors influencing variation in basal metabolic rate include fat-free mass, fat mass, age, and circulating thyroxine but not sex, circulating leptin, or triiodothyronine'. *American Journal of Clinical Nutrition.* 82(5), 941–48.

Josefsson, T., Lindwall, M., & Archer, T. (2014) 'Physical exercise intervention in depressive disorders: meta-analysis and systematic review'. *Scandinavian Journal of Medicine & Science in Sports.* 24(2), 259–72.

Kang, S., Kyung, C., Park, J.S., Kim, S., Lee, S., Kim, M.K., Kim, H.K., Kim, K.R., Jeon, T.J., & Ahn, C.W. (2014) 'Subclinical vascular inflammation in subjects with normal weight obesity and its association with body fat: an F-FDG-PET/CT study'. *Cardiovascular Diabetology.* 13(70)

Karremans, J.C., Frankenhuis, W.E., & Arons, S. (2010) 'Blind men prefer a low waist-to-hip ratio'. *Evolution & Human Behavior.* 31(3), 182–86.

Keel, P.K. & Klump, K.L (2003) 'Are eating disorders culture-bound syndromes? Implications for conceptualizing their etiology'. *Psychological Bulletin.* 129(5), 747–69.

Kerwin, D.R., Zhang, Y., Kotchen, J.M., Espeland, M.A., Van Horn, L., McTigue, K.M., Robinson, J.G., Powell, L., Kooperberg, C., Coker, L.H., & Hoffmann, R. (2010) 'The cross-sectional relationship between body mass index, waist–hip ratio, and cognitive performance in postmenopausal women enrolled in the women's health initiative'. *Journal of the American Geriatrics Society.* 58(8), 1427–32.

Kimura, K., Ozeki, M., Juneja, L., & Ohira, H. (2007) 'L-Theanine reduces psychological and physiological stress responses'. *Biological Psychology.* 74(1), 39–45.

King, N.A. & Blundell, J.E. (1995) 'High-fat foods overcome the energy expenditure induced by high-intensity cycling or running'. *European Journal of Clinical Nutrition.* 49(2), 114–23.

Kitahara, C.M., Flint, A.J., Berrington de Gonzalez, A., Bernstein, L., Brotzman, M., MacInnis, R.J., Moore, S.C., Robien, K., Rosenberg, P.S., Singh, P.M., Weiderpass, E., Adami, H.O., et al. (2014) 'Association between class III obesity (BMI of 40–59 kg/m^2) and mortality: a pooled analysis of 20 prospective studies'. *PLOS One.* DOI: 10.1371/ journal. pmed.1001673

Kort, J.D., Winget, C., Kim, S.H., & Lathi, R.B. (2014) 'A retrospective cohort study to evaluate the impact of meaningful weight loss on fertility outcomes in an overweight population with infertility'. *Fertility and Sterility.* 101(5), 1400–03.

Kramer, C.K., Zinman, B., & Retnakaran, R. (2013) 'Are metabolically healthy overweight and obesity benign conditions?: a systematic review and meta-analysis'. *Annals of Internal Medicine.* 159(11), 758–69.

Kristensen, J., Vestergaard, M., Wisborg, K., Kesmodel, U., & Secher, N.J. (2005) 'Pre-pregnancy weight and the risk of stillbirth and neonatal death'. *International Journal of Obstetrics & Gynaecology.* 112(4), 403–08.

Kroke, A., Klipstein-Grobusch, K., Voss, S., Mosender, J., Thielecke, F., Noack, R., & Boeing, H. (1999) 'Validation of a self-administered food-frequency questionnaire administered in the European prospective investigation into cancer and nutrition (EPIC) study: comparison of energy, protein, and macronutrient intakes estimated with the doubly-labeled water, urinary nitrogen, and repeated 24-h dietary recall methods'. *American Journal of Clinical Nutrition.* 70, 439–47.

Kuhnle, G.C., Tasevska, N., Lentjes, M., Griffin, J.L., Sims, M.A., Richardson, L., Aspinall, S.M., Mulligan, A.A., Luben, R.N., & Khaw, K-K. (2015) 'Association between sucrose intake and risk of overweight and obesity in a prospective sub-cohort of the European Prospective Investigation into Cancer in Norfolk (EPIC-Norfolk)'. *Public Health Nutrition.* DOI: dx.doi.org/10.1017/S1368980015000300

Ledochowski, L., Ruedl, G., Taylor, A.H., & Kopp, M. (2015). 'Acute effects of brisk walking on sugary snack cravings in overweight people, affect and responses to a manipulated stress situation and to a sugary snack cue: a crossover study'. *PLOS One*. Online at: journals.plos.org/plosone/article?id=10.1371/journal.pone.0119278

Lichtman, S.W., Pisarska, K., Berman, E.R., Pestone, M., Dowling, H., Offenbacher, E., Weisel, H., Heshka, S., Matthews, D.E., & Heymsfield, S.B. (1992) 'Discrepancy between self-reported and actual caloric intake and exercise in obese subjects'. *The New England Journal of Medicine*. 327 (37), 1893–98.

Live Science (2011) '5 health benefits of smoking'. Online at: www.livescience.com/15115-5-health-benefits-smoking-disease.html

Locke, E.A. & Latham, G.P. (1990). *A Theory of Goal-Setting and Task Performance*. Englewood Cliffs, NJ: Prentice Hall.

Locke, E.A. & Latham, G.P. (2002). 'Building a Practically Useful Theory of Goal Setting and Task Motivation'. *American Psychologist*. 57(9), 705–17.

Mangner, N., Scheuermann, K., Winzer, E., Wagner, I., Hoellriegel, R., Sandri, M., Zimmer, M., Mende, M., Linke, A., Kiess, W., Schuler, G., Körner, A., & Erbs, S. (2014) 'Childhood obesity: impact on cardiac geometry and function'. *Journal of the American College of Cardiology*. 7(12), 1198–205.

Manninen, A.H. (2004) 'High-protein weight loss diets and purported adverse effects: where is the evidence?'. *Journal of the International Society of Sports Nutrition*. 1, 45–51.

Männistö, S., Harald, K., Kontto, J., Lahti-Koski, M., Kaartinen, N.E., Saarni, S.E., Kanerva, N., & Jousilahti, P. (2014) 'Dietary and lifestyle characteristics associated with normal-weight obesity: the National FINRISK 2007 Study'. *British Journal of Nutrition*. 111(5), 887–94.

Manson, J.E., Willett, W.C., Stampfer, M.J., Colditz, G.A., Hunter, D.J., Hankinson, S.E., Hennekens, C.H., & Speizer, F.E. (1995) 'Body weight and mortality among women'. *New England Journal of Medicine*. 333, 677–85.

Mason, C., Foster-Schubert, K.E., Imayama, I., Xiao, L., Kong, A., Campbell, K.L., Duggan, C.R., Wang, C., Alfano, C.M., Ulrich, C.M.,

Blackburn, G.L., & McTiernan, A. (2013) 'History of weight cycling does not impede future weight loss or metabolic improvements in post-menopausal women'. *Metabolism.* 62(1), 127–36.

Masoro, E.J. (2005) 'Dietary restriction, longevity and ageing — the current state of our knowledge and ignorance'. *Mechanisms of Ageing and Development.* 126(9), 913–22.

Masters, R.K., Reither, E.N., Powers, D.A., Yang, Y.C., Burger, A.E., & Link, B.G. (2013) 'The impact of obesity on US mortality levels: the importance of age and cohort factors in population estimates'. *American Journal of Public Health.* 103(10), 1895–901.

McDonald, L. (2017) 'Muscle gain math'. Online at: bodyrecomposition. com/muscle-gain/muscle-gain-math.html/

McGuire, M., Wing, R., & Hill, J. (1999) 'The prevalence of weight loss maintenance among American adults'. *International Journal of Obesity* (23), 1314–19.

Medical Daily (2015) 'Texas doctors diagnose 3½ year-old toddler with type 2 diabetes'. Online at: www.medicaldaily.com/texas-doctors-diagnose-312-year-old-toddler-type-2-diabetes-352980

MedicineNet (2015) 'Hypoglycemia (low blood sugar) symptoms and diabetes'. Online at: www.medicinenet.com/hypoglycemia_low_blood_sugar_symptoms_and_diabetes/views.htm

Mehta, T., Smith Jr., D.L., Muhammad, J., & Casazza, K. (2014) 'Impact of weight cycling on risk of morbidity and mortality'. *Obesity Reviews.* 15(11), 870–81.

Melissas, J., Volakakis, E., & Hadjipavlou, A. (2003) 'Low-back pain in morbidly obese patients and the effect of weight loss following surgery'. *Obesity Surgery.* 13(3), 389–93.

Menke, A., Casagrande, S., Geiss, L., & Cowie, C.C. (2015) 'Prevalence of and trends in diabetes among adults in the United States, 1988–2012'. *JAMA.* 314(10), 1021–29.

Mensink, G.B.M., Schienkiewitz, A., Haftenberger, N., Lampert, T., Ziese, T., & Scheidt-Nave, C. (2013) 'Übergewicht und Adipositas in Deutschland'. *Bundesgesundheitblatt* 56, 786–94.

Merino, J., Megias-Rangil, I., Ferré, R., Plana, N., Girona, J., Rabasa, A.,

Aragonés, G., Cabré, A., Bonada, A., Heras, M., & Masana, L. (2012) 'Body weight loss by very-low-calorie diet program improves small artery reactive hyperemia in severely obese patients'. *Obesity Surgery*. 23(1), 17–23.

Moffatt, R.J. & Owens, S.G. (1991) 'Cessation from cigarette smoking: changes in body weight, body composition, resting metabolism, and energy consumption'. *Metabolism*. 40(5), 465–70.

Mokdad, A.H., Marks, J.S., Stroup, D.F., & Gerberding, J.L. (2004) 'Actual causes of death in the United States, 2000'. *Journal of the American Medical Association*. 291(10), 1238–45.

Muhlheim, L.S., Allison, D.B., Heshka, S., & Heymsfield, S.B. (1998) 'Do unsuccessful dieters intentionally underreport food intake?' *The International Journal of Eating Disorders*. 24(3), 259–66.

Muraven, M., Tice, D.M., & Baumeister, R.F. (1998) 'Self-control as a limited resource: regulatory depletion patterns'. *Journal of Personality and Social Psychology*. 74(3), 774–89.

Nationale Arbeitsgruppe zur Prävention und Behandlung von Adipositas (1994) 'Weight cycling. National Task Force on the Prevention and Treatment of Obesity'. *Journal of the American Medical Association*. 272(15), 1196–202.

National Post (2015) 'Nearly one in five severely obese Canadians die in hospital after an emergency surgery: study'. Online at: news.nationalpost.com/health/nearly-one-in-five-severely-obese-canadians-die-in-hospital-after-an-emergency-surgery

NBC News. (2011) 'Yo-yo dieting better than staying obese. Losing and gaining weight is healthier than no weight loss at all, study finds'. Online at: www.nbcnews.com/id/43298247/ns/health-diet_and_nutrition/t/yo-yo-dieting-better-staying-obese/#.VO1jvvnF-T8

Neal, D., Wood, W., Wu, M., & Kurlander, D. (2011). 'The pull of the past: when do habits persist despite conflict with motives?' *Personality and Social Psychology Bulletin*. 37(11) 1428–37.

Nelson, M.E., Fiatarone, M.A., Morganti, C.M., Trice, I., Greenberg, R.A., & Evans, W.J. (1994) 'Effects of high-intensity strength training on multiple risk factors for osteoporotic fractures: a randomized controlled trial'. *Journal of the American Medical Association*. 272(24), 1909–14.

NewsMic (2014) 'Anyone silly enough to think fat is good for you needs to see this brain study'. Online at: mic.com/articles/92549/anyone-silly-enough-to-think-fat-is-good-for-you-needs-to-see-this-brain-study?utm_source=takepart&utm_ medium=social&utm_ campaign=July

New York Times (2012) 'Are most people in denial about their weight?' Online at: well.blogs.nytimes.com/2012/04/18/are-most-people-in-denial-about-their-weight/?_r=0

Nickols-Richardson, S.M., Coleman, M.D., Volpe, J.J., & Hosig, K.W. (2005) 'Perceived hunger is lower and weight loss is greater in overweight premenopausal women consuming a low-carbohydrate/high-protein vs high-carbohydrate/low-fat diet'. *Journal of the American Dietetic Association.* 105(9), 1433–37.

Niedermaier, T., Behrens, G., Schmid, D., Schlecht, I., Fischer, B., & Leitzmann, M.F. (2015) 'Body mass index, physical activity, and risk of adult meningioma and glioma. A meta-analysis'. *Neurology.* 85(15), 1342–50. DOI: 10.1212/WNL.0000000000002020

Nielsen, S.J. & Popkin, B.M. (2003) 'Patterns and trends in food portion sizes, 1977–1998'. *Journal of the American Medical Association.* 289(4), 450–53.

Norman, R.J., Noakes, M., Wu, R., Davies, M.J., Moran, L., & Wang, J.X. (2004) 'Improving reproductive performance in overweight/obese women with effective weight management'. *Human Reproduction Update.* 10(3), 267–80.

Oettingen, G. (1996) *The Psychology of Action: linking cognition and motivation to behavior.* New York: The Guilford Press.

Oh, H. & Taylor, A.H. (2012) 'Brisk walking reduces ad libitum snacking in regular chocolate eaters during a workplace simulation'. *Appetite.* 58(1), 387–92.

Okorodudu, D.O., Jumean, M.F., Montori, V.M., Romero-Corral, A., Somers, V.K., Erwin, P.J., & Lopez-Jimenez, F. (2010) 'Diagnostic performance of body mass index to identify obesity as defined by body adiposity: a systematic review and meta-analysis'. *International Journal of Obesity.* 34(5), 791–99.

Oliveros, E., Somers, V.K., Sochor, O., Goel, K., & Lopez-Jimenez, F. (2014) 'The concept of normal weight obesity'. *Progress in Cardiovascular Diseases*. 56(4), 426–33.

Onyike, C.U., Crum, R.M., Lee, H.B., Lyketsos, C.G., & Eaton, W.W. (2003) 'Is obesity associated with major depression? Results from the Third National Health and Nutrition Examination Survey'. *American Journal of Epidemiology*. 158(12), 1139–47.

Oppenheimer, R., Howells, K., Palmer, R.L., & Chaloner, D.A. (1985) 'Adverse sexual experience in childhood and clinical eating disorders: a preliminary description'. *Journal of Psychiatric Research*. 19(2/3), 357–61.

Oreopoulos A., Padwal, R., Kalantar-Zadeh, K., Fonarow, G.C., Norris, C.M., & McAlister, F.A. (2008) 'Body mass index and mortality in heart failure: a meta-analysis'. *American Heart Journal*. 156(1), 13–22.

Ou, X., Andres, A., Pivik, R.T., Cleves, M.A., & Badger, T.M. (2015a) 'Brain gray and white matter differences in healthy normal weight and obese children'. *Journal of Magnetic Resonance Imaging*. DOI: 10.1002/jmri.24912.

Ou, X., Thakali, K.M., Shankar, K., Andres, A., & Badger, T.M. (2015b) 'Maternal adiposity negatively influences infant brain white matter development'. *Obesity*. 23(5), 1047–54.

Pacanowski, C.R. & Levitsky, D.A. (2015) 'Frequent self-weighing and visual feedback for weight loss in overweight adults'. *Journal of Obesity*. DOI:10.1155/2015/763680

Parent, M.C. & Alquist, J.L. (2015) 'Born fat: the relations between weight changeability beliefs and health behaviors and physical health'. *Health Education and Behavior*. Online at: heb.sagepub.com/content/early/2015/08/25/1090198115602266.full

Park, H.J., Lee, S.E., Kim, H.B., Isaacson, R.E., Seo, K.W., & Song, K.H. (2015) 'Association of obesity with serum leptin, adiponectin, and serotonin and gut microflora in beagle dogs'. *Journal of Veterinary Internal Medicine*. 29(1), 43–50.

Park, M. (2010) 'Twinkie diet helps nutrition professor lose 27 pounds'. *CNN*. Online at: edition.cnn.com/2010/HEALTH/11/08/twinkie.diet.professor/index.html

Paul, T.K., Sciacca, R.R., Bier, M., Rodriguez, J., Song, S., & Giardina, E.V. (2014) 'Size misperception among overweight and obese families'. *Journal of General Internal Medicine*. 30(1), 43–50.

Pawlowski, A. (2014) 'Man loses nearly 40 lbs. eating only McDonald's'. *Today*. Online at: www.today.com/health/man-loses-nearly-40-lbs-eating-only-mcdonalds-2D11863528

Pellegrinelli, V., Rouault, C., Rodriguez-Cuenca, S., Albert, V., Edom-Vovard, F., Vidal-Puig, A., Clément, K., Butler-Browne, G.S., & Lacasa, D. (2015) 'Human adipocytes induce inflammation and atrophy in muscle cells during obesity'. *Diabetes*. 64(9), 3121–34. DOI: 10.2337/db14-0796

Pemberton, M. (2014) 'As a doctor, I'd rather have HIV than diabetes'. *The Spectator*. Online at: www.spectator.co.uk/2014/04/why-id-rather-have-hiv-than-diabetes/

Perkins, K.A., Epstein, L.H., Stiller, R.L., Marks, B.L., & Jacob, R.G. (1989) 'Acute effects of nicotine on resting metabolic rate'. *American Journal of Clinical Nutrition*. 50, 545–50.

Pesheva, E. (2014) 'Study: obesity fuels silent heart damage, increased risk of future heart failure. Evidence of heart muscle damage seen even among symptom-free people'. John Hopkins News Network: 25 November. Online at: hub.jhu.edu/2014/11/25/obesity-heart-disease-risk

Pietiläinen, K.H., Korkeila, M., Bogl, L.H., Westerterp, K.R., Yki-Järvinen, H., Kaprio, J., & Rissanen, A. (2010) 'Inaccuracies in food and physical activity diaries of obese subjects: complementary evidence from doubly labeled water and co-twin assessments'. *International Journal of Obesity*. 34(3), 437–45.

Psychology Today (2014) 'Eternal Curves'. Online at: www.psychologytoday.com/articles/201206/eternal-curves

Puhl, R.M., Andreyeva, T., & Brownell, K.D. (2008) 'Perceptions of weight discrimination: prevalence and comparison to race and gender discrimination in America'. *International Journal of Obesity*. 32, 992–1000.

Purcell, K., Sumithran, P., Prendergast, L.A., Bouniu, C.J., Delbridge, E., & Proietto, J. (2014) 'The rate of weight loss does not influence long term weight maintenance: a randomised controlled trial'. *Obesity Research & Clinical Practice*. 8(1), 80.

Quartz (2015) 'Why you can't just eat one french fry (or donut or slice of pizza)'. Online at: qz.com/447305/why-you-cant-just-eat-one-french-fry-or-piece-of-pizza-or-donut/

Rathmann, W., Haastert, B., Icks, A., Löwel, H., Meisinger, C., Holle, R., & Giani, G. (2003) 'High prevalence of undiagnosed diabetes mellitus in Southern Germany: target populations for efficient screening. The KORA survey 2000'. *Diabetologia.* 46(2), 182–89.

Redman, L.M., Heilbronn, L.K., Martin, C.K., de Jonge, L., Williamson, D.A., Delany, J.P., & Ravussin, E. (2009) 'Metabolic and behavioral compensations in response to caloric restriction: implications for the maintenance of weight loss'. Online at: journals.plos.org/plosone/article?id=10.1371/journal.pone.0004377#pone-0004377-g004

Rich-Edwards, J.W., Spiegelman, D., Garland, M., Hertzmark, E., Hunter, D.J., Colditz, G.A., Willett, W.C., Wand, H., & Manson, J.E. (2002) 'Physical activity, body mass index, and ovulatory disorder infertility'. *Epidemiology.* 13(2), 184–90.

Robinson, E., Parretti, H., & Aveyard, P. (2014) 'Visual identification of obesity by healthcare professionals: an experimental study of trainee and qualified GPs'. *British Journal of General Practice.* Online at: bjgp.org/content/64/628/e703.full.pdf+html?

Rolls, B.J., Rowe, E.A., Rolls, E.T., Kingston, B., Megson, A., & Gunary, R. (1981) 'Variety in a meal enhances food intake in man'. *Physiology & Behavior.* 26(2), 215–21.

Romero-Corral, A., Montori, V.M., Somers, V.K., Korinek, J., Thomas, R.J., Allison, T.G., Mookadam, F., & Lopez-Jimenez, F. (2006) 'Association of bodyweight with total mortality and with cardiovascular events in coronary artery disease: a systematic review of cohort studies'. *Lancet.* 368(9536), 666–78.

Rosenbaum, M., Hirsch, J., Gallagher, D.A., & Leibel, R.L. (2008) 'Long-term persistence of adaptive thermogenesis in subjects who have maintained a reduced body weight'. *American Journal of Clinical Nutrition.* 88(4), 906–12.

Saag, K.G. & Choi, H. (2006) 'Epidemiology, risk factors, and lifestyle modifications for gout'. *Arthritis Research and Therapy.* Online at: www.

biomedcentral.com/content/pdf/ar1907.pdf

Sandholt, C.H., Hansen, T., & Pedersen, O. (2012) 'Beyond the fourth wave of genome-wide obesity association studies'. *Nutrition and Diabetes.* 2, e37. DOI:10.1038/nutd.2012.9

Sarafrazi, N., Hughes, J.P., Borrud, L., Burt, V., & Paulose-Ram, R. (2014) 'Perception of weight status in U.S. children and adolescents aged 8–15 Years, 2005–2012'. Online at: www.cdc.gov/nchs/data/databriefs/db158. htm

Saris, W.H.M. (2012) 'Very-low-calorie diets and sustained weight loss'. *Obesity Research Special Issue: Dietary Patterns for Weight Management and Health.* 9(11), 295S–301S.

Sarna, S., Sahi, T., Koskenvuo, M., & Kaprio, J. (1993) 'Increased life expectancy of world class male athletes'. *Medicine and Science in Sports and Exercise.* 25(2), 237–44.

Sawyer B.J., Bhammar, D.M., Angadi, S.S., Ryan, D.M., Ryder, J.R., Sussman, E.J., Bertmann, F.M., & Gaesser, G.A. (2015) 'Predictors of fat mass changes in response to aerobic exercise training in women'. *Journal of Strength and Conditioning Research.* 29(2), 297–304.

Schoeller, D.A. (2014) 'Eating patterns, diet quality and energy balance. The effect of holiday weight gain on body weight'. *Physiology & Behavior.* 134, 66–69.

Schwartz, A.R., Gold, A.R., Schubert, N., Stryzak, A.,Wise, R.A., Permutt, S., & Smith, P.L. (1991) 'Effect of weight loss on upper airway collapsibility in obstructive sleep apnea'. *American Review of Respiratory Disease.* 144(3), 494–98.

ScienceDaily (2012) 'Obese youth have significantly higher risk of gallstones'. Online at: www.sciencedaily.com/releases/2012/08/120824205733.htm

ScienceDaily (2014) 'Study shows keys to successful long-term weight loss maintenance'. Online at: www.sciencedaily.com/releases/2014/01/140106115351.htm

ScienceDaily (2014). 'Losing weight won't necessarily make you happy, researchers say'. Online at: www.sciencedaily.com/releases/2014/08/140807105430.htm

Seo, B.R., Bhardwaj, P., Choi, S., Gonzalez, J., Andresen Eguiluz, R.C.,

Wang, K., Mohanan, S., Morris, P.G., Du, B., Zhou, X.K., Vahdat, L.T., Verma, A., Elemento, O., Hudis, C.A., Williams, R.M., Gourdon, D., Dannenberg, A.J., & Fischbach, C. (2015) 'Obesity-dependent changes in interstitial ECM mechanics promote breast tumorigenesis'. *Science Translational Medicine*, 7(301), 301ra130.

Shah, N.R. & Braverman, E.R. (2012) Measuring adiposity in patients: the utility of body mass index (BMI), percent body fat, and leptin'. *PLOS One*. 7(4), e33308.

Shammas, M.A. (2012) 'Telomeres, lifestyle, cancer, and aging'. *Current Opinion in Clinical Nutrition and Metabolic Care*. 14(1), 28–34.

Skorka-Brown, J., Andrade, J., & May, J. (2014) 'Playing "Tetris" reduces the strength, frequency and vividness of naturally occurring cravings'. *Appetite*. (76), 161–65.

Simopoulos, A.P. (2005) *Nutrition and Fitness*. Washington, D.C.: Karger.

Smith, P.L., Gold, A.R., Meyers, D.A., Haponik, E.F., & Bleecker, E. (1985) 'Weight loss in mildly to moderately obese patients with obstructive sleep apnea'. *Annals of Internal Medicine*. 103(6), 850–55.

Spectator (2013) 'Obesity is not a disease'. Online at: www.spectator.co.uk/features/9049971/the-battle-of-the-bulge/

Spectator (2014) 'As a doctor, I'd rather have HIV than diabetes'. Online at: www.spectator.co.uk/features/9185591/why-id-rather-have-hiv-than-diabetes/

Spencer, S.J., Steele, C.M., & Quinn, D.M. (1999) 'Stereotype threat and women's math performance'. *Journal of Experimental Social Psychology*. 35(1), 4–28.

Spiegel (2012) 'Ernährung: Schokolade-Liebhaber sind schlanker'. Online at: www.spiegel.de/gesundheit/ernaehrung/ernaehrung-wer-oft-schokolade-isst-ist-duenner-a-836016.html

Spring, B., Howe, D., Berendsen, M., McFadden, H.G., Hitchcock, K., Rademaker, A.W., & Hitsman, B. (2009) 'Behavioral intervention to promote smoking cessation and prevent weight gain: a systematic review and meta-analysis'. *Addiction*. 104(9), 1472–86.

Stamford, B.A., Matter, S., Fell, R.D., & Papanek, P. (1986) 'Effects of smoking cessation on weight gain, metabolic rate, caloric consumption, and blood lipids'. *American Journal of Clinical Nutrition.* 43(4) 486–94.

Stein, P.D., Beemath, A., & Olson, R.E. (2005) 'Obesity as a risk factor in venous thromboembolism'. *The American Journal of Medicine.* 118(9), 978–80.

Steinberg, D.M., Bennett, G.G., Askew, S., & Tate, D.F. (2015) 'Weighing every day matters: daily weighing improves weight loss and adoption of weight control behaviors'. *Journal of the Academy of Nutrition and Dietetics.* 115(4), 511–18.

Stevens, V.L, Jacobs, E.J., Sun, J., Patel, A.V., McCullough, M.L., Teras, L.R., & Gapstur, S.M. (2012) 'Weight cycling and mortality in a large prospective US study'. *American Journal of Epidemiology.* 175(8), 785–92.

Stewart, W.K. & Fleming, L.W. (1973) 'Features of a successful therapeutic fast of 382 days' duration'. *Postgraduate Medical Journal.* 49, 203–09.

Stirrat, L.I., & Reynolds, R.M. (2014). 'Effects of maternal obesity on early and long-term outcomes for offspring'. *Research and Reports in Neonatology.* 4, 43–53.

Stock, S.A.K., Redaelli, M., Wendland, G., Civello, D., & Lauterbach, K.W. (2005) 'Diabetes-prevalence and cost of illness in Germany: a study evaluating data from the statutory health insurance in Germany'. *Diabetic Medicine.* 23(3), 299–305.

Strain, G.W., Kolotkin, R.L., Dakin, G.F., Gagner, M., Inabnet, W.B., Christos, P., Saif, T., Crosby, R., & Pomp, A. (2014) 'The effects of weight loss after bariatric surgery on health-related quality of life and depression'. *Nutrition & Diabetes.* 4, e132.

Subak, L.L., Johnson, C., Johnson, C., Whitcomb, E., Boban, D., Saxton, J., & Brown, J.S. (2002) 'Does weight loss improve incontinence in moderately obese women?'. *International Urogynecology Journal.* 13(1), 40–43.

Subak, L.L., Whitcomb, E., Shen, H., Saxton, J., Vittinghoff, E., & Brown, J.S. (2005) 'Weight loss: a novel and effective treatment for urinary incontinence'. *The Journal of Urology.* 174(1), 190–95.

Surwit, R.S., Feinglos, M.N., McCaskill, C.C., Clay, S.L., Babyak, M.A., Brownlow, B.S., Plaisted, C.S., & Lin, P.H. (1997) 'Metabolic and behavioral effects of a high-sucrose diet during weight loss'. *The American Journal of Clinical Nutrition.* 65(4), 908–15.

Sweet, L. (2014) 'Fantasy bodies, imagined pasts: a critical analysis of the "Rubenesque" fat body in contemporary culture'. *Fat Studies: An Interdisciplinary Journal of Body Weight and Society.* 3(2), 130–42.

Sydney Morning Herald (2015) '"It's wrong to tell fat women they look fabulous"' Online at: www.smh.com.au/lifestyle/diet-and-fitness/its-wrong-to-tell-fat-women-they-look-fabulous-20150423-1ms5r0. html?stb=red

Tanda, R., Salsberry, P.J., Reagan, P.B., & Fang, M.Z. (2012) 'The impact of prepregnancy obesity on children's cognitive test scores'. *Maternal and Child Health Journal.* 17(2), 222–29.

Telegraph (2015) 'Don't worry — 20 minutes exercise a week is enough, say experts'. Online at: www.telegraph.co.uk/news/science/science-news/11360617/NHS-weekly-exercise-guidelines-are-too-demanding-say-health-experts.html

Thomas, J.G., Bond, D.S., Phelan, S., Hill, J.O., & Wing, R.R. (2014) 'Weight-loss maintenance for 10 years in the National Weight Control Registry'. *American Journal of Preventive Medicine.* 46(1), 17–23.

Today (2014) 'Man loses 56 pounds after eating only McDonald's for six months'. Online at: www.today.com/health/man-loses-56-pounds-after-eating-only-mcdonalds-six-months-2D79329158

Toubro, S., Sørensen, T.I., Rønn, B., Christensen, N.J., & Astrup, A. (2013) 'Twenty-four-hour energy expenditure: the role of body composition, thyroid status, sympathetic activity, and family membership'. *The Journal of Clinical Endocrinology & Metabolism.* 81(7), DOI: dx.doi.org/10.1210/jcem.81.7.8675595

Truhler, K. (2012) 'Sizing up old Hollywood'. *GlamAmor.* Online at: www.glamamor.com/2013/08/sizing-up-old-hollywood-heights-weights.html

Tsiros, M., Buckley, J., Howe, P., Walkley, J., Hills, A., & Coates, A. (2014) 'Musculoskeletal pain in obese compared with healthy-weight children'. *Clinical Journal of Pain.* 30(7), 583–88.

Turner, L.R., Harris, M.F., & Mazza, D. (2014) 'Obesity management in general practice: does current practice match guideline recommendations?'. *The Medical Journal of Australia.* Online at: www.mja.com.au/journal/2015/202/7/obesity-management-general-practice-does-current-practice-match-guideline

TZ (2012) 'Keine Angst vor der letzten Zigarette'. Online at: www.tz.de/leben/gesundheit/klappt-gute-vorsatz-rauchen-aufzuhoeren-2669707.html

Tzankoff, S.P. & Norris, A.H. (1977) 'Effect of muscle mass decrease on age-related BMR changes'. *Journal of Applied Physiology*. 43(6), 1001–06.

Uni Göttingen (2005) 'Was hat Bluthochdruck mit dem Gewicht zu tun?' Online at: www.med.uni-goettingen.de/de/media/tag_der_medizin/tdm2005_was_hat_bluthochd_m_gewicht.pdf

Upton, J. (2008) 'Global Metabolism Myths'. Online at: www.globalhealthandfitness.com/metabolism%20myths.htm

Urban, L.E., Dallal, G.E., Robinson, L.M., Ausman, L.M., Saltzman, E., & Roberts, S.B. (2010) 'The accuracy of stated energy contents of reduced-energy, commercially prepared foods'. *Journal of The American Dietetic Association*. 110(1), 116–23.

Vaughan, L., Zurlo, F., & Ravussin, E. (1991) 'Aging and energy expenditure'. *The American Journal of Clinical Nutrition*. 53(4), 821–25.

Verreijen, A.M., Verlaan, S., Engberink, M.F., Swinkels, S., de Vogel-van den Bosch, J., & Weijs, P.J. (2014) 'A high whey protein-, leucine-, and vitamin D-enriched supplement preserves muscle mass during intentional weight loss in obese older adults: a double-blind randomized controlled trial'. *The American Journal of Clinical Nutrition*. 101(2), 279–86.

Wadden, T.A., Neiberg, R.H., Wing, R.R., Clark, J.M., Delahanty, L.M., Hill, J.O., Krakoff, J., Otto, A., Ryan, D.H., Vitolins, M.Z., & The Look AHEAD Research Group (2011) 'Four-year weight losses in the Look AHEAD Study: factors associated with long-term success'. *Obesity*. 19(10), 1987–98.

Wang, G., Djafarian, K., Egedigwe, C.A., El Hamdouchi, A., Ojiambo, R., Ramuth, H., Wallner-Liebmann, S.J., Lackner, S., Diouf, A., Sauciuvenaite, J., Hambly, C., Vaanholt, L.M., Faries, M.D., & Speakman, J.R. (2015) 'The relationship of female physical attractiveness to body fatness'. *PeerJ*. 3(e1155). dx.doi.org/10.7717/peerj.1155.

The Washington Post (2014) 'Parents of obese kids aren't always willing to admit their child has a health problem'. Online at: www.washingtonpost. com/news/to-your-health/wp/2014/07/25/parents-of-obese-kids-arent-always-willing-to-admit-their-child-has-a-health-problem/

The Washington Post (2015) 'The science of skipping breakfast: how government nutritionists may have gotten it wrong'. Online at: www. washingtonpost.com/news/wonkblog/wp/2015/08/10/the-science-of-skipping-breakfast-how-government-nutritionists-may-have-gotten-it-wrong/

Watters, E. (2006) 'DNA is not destiny'. *Discover.* 27(11), 32–75.

WebMD (2006) 'Extreme obesity in tots tied to low IQ — difference of 25–30 points seen in IQs of those very obese by age 4'. Online at: www. webmd.com/parenting/news/20060831/extreme-obesity-in-tots-tied-to-low-iq?page=2

Weigle, D.S., Breen, P.A., Matthys, C.C., Callahan, H.S., Meeuws, K.E., Burden, V.R., & Purnell, J.Q. (2005) 'A high-protein diet induces sustained reductions in appetite, ad libitum caloric intake, and body weight despite compensatory changes in diurnal plasma leptin and ghrelin concentrations'. *American Journal of Clinical Nutrition.* 82(1), 41–48.

Weinsier, R., Hunter, G., Zuckerman, P., Redden, D., Darnell, B., & Larson, D. (2000) 'Energy expenditure and free-living physical activity in black and white women: comparison and after weight loss'. *American Journal of Clinical Nutrition.* 71, 1138–46.

Willett, W.C., Manson, J.E., Stampfer, M.J., Colditz, G.A., Rosner, B., Speizer, F.E., & Hennekens, C.H. (1995) 'Weight, weight change, and coronary heart disease in women: risk within the "normal" weight range'. *Journal of the American Medical Association.* 273(6), 461–65.

Wilson, R.M., Marshall, N.E., Jeske, D.R., Purnell, J.Q., Thornburg, K., & Messaoudi, I. (2015) 'Maternal obesity alters immune cell frequencies and responses in umbilical cord blood samples'. *Pediatric Allergy and Immunology.* 26(4), 344–51.

Wing, R.R. & Phelan, S. (2005) 'Long-term weight loss maintenance'. *The American Journal of Clinical Nutrition.* 82(1), 222S–25S.

Wood, P.D., Stefanick, M.L., Dreon, D.M., Frey-Hewitt, B., Garay, S.C., Williams, P.T., Superko, R., Fortmann, S.P., Albers, J.J., Vranizan,

K.M., Ellsworth, N.M., Terry, R.B., & Haskell, W.L. (1988) 'Changes in plasma lipids and lipoproteins in overweight men during weight loss through dieting as compared with exercise'. *New England Journal of Medicine.* 319, 1173–79.

Wright, D.A., Sherman, W.M., & Dernbach, A.R. (1991) 'Carbohydrate feedings before, during, or in combination improve cycling endurance performance'. *Journal of Applied Physiology.* 71(3), 1082–88.

Xu, C., Cai, Y., Fan, P., Bai, B., Chen, J., Deng, H.B., Che, C.M., Xu, A., Vanhoutte, P.M., & Wang, Y. (2015) 'Calorie restriction prevents metabolic aging caused by abnormal SIRT1 function in adipose tissues'. *Diabetes.* 64(5), 1576–90.

Zauner, C., Schneeweiss, B., Kranz, A., Madl, C., Ratheiser, K., Kramer, L., Roth, E., Schneider, B., & Lenz, K. (2000) 'Resting energy expenditure in short-term starvation is increased as a result of an increase in serum norepinephrine'. *The American Journal of Clinical Nutrition.* 71(6), 1511–15.

Die Zeit (2008) 'Stimmt's: Süßstoff für Ferkel'. Online at: www.zeit.de/2008/15/Stimmts-Suessstoff

Die Zeit (2012) 'Die fette Gefahr'. Online at: www.zeit.de/2012/31/Uebergewicht-Krebs